Full-Stack Web Development with GraphQL and React

Second Edition

Taking React from frontend to full-stack with
GraphQL and Apollo

Sebastian Grebe

BIRMINGHAM—MUMBAI

Full-Stack Web Development with GraphQL and React
Second Edition

Group Product Manager: Pavan Ramchandani
Publishing Product Manager: Kaustubh Manglurkar
Senior Editor: Keagan Carneiro
Content Development Editor: Adrija Mitra
Technical Editor: Saurabh Kadave
Copy Editor: Safis Editing
Project Coordinator: Rashika Ba
Proofreader: Safis Editing
Indexer: Pratik Shirodkar
Production Designer: Aparna Bhagat
Marketing Coordinator: Anamika Singh

First published: January 2019

Second edition: March 2022

Production reference: 1280122

Published by Packt Publishing Ltd.
Livery Place
35 Livery Street
Birmingham
B3 2PB, UK.
ISBN: 978-1-80107-788-0

www.packt.com

I thank my love, who has been able to give me the time I needed. Her support made every day better and her help accomplishing my goals is just great!

– Sebastian Grebe

Contributors

About the author

Sebastian Grebe is a verified computer science expert for application development. He is a young entrepreneur working on a variety of products targeting the consumer market. He specializes in web development using the newest technologies, such as React, the Phoenix framework, Kubernetes, and many more. Furthermore, he has experience in merging old and new applications, developing cross-platform apps with React Native, and writing efficient APIs and backends with Node.js and Elixir.

Currently, he works as an engineering manager on a microservice-oriented architecture using micro frontends to power a scalable e-commerce platform.

About the reviewer

Devlin Basilan Duldulao is a full-stack engineer with over 8 years of web, mobile, and cloud development experience. He has been a recipient of Microsoft's **Most Valuable Professional (MVP)** award since 2018 and earned the title of Auth0 Ambassador for his passion for sharing best practices in application security. Devlin has passed some prestigious software and cloud development exams, such as MSCD, Azure Associate Developer, AWS Associate Developer, and Terraform Associate.

He has also authored the book *ASP.NET Core and Vue.js* and coauthored the book *Practical Enterprise React* in 2020 and 2021, amid the pandemic.

I would like to thank the whole Packt team for trusting me to review this book and assisting in the process along the way. I would like to thank my wife, Ruby Jane Cabagnot, a full-stack developer, for sharing her thoughts about the book. I would like to thank my parents, Lucy and Alberto, and my in-laws, Ruben and Nitz, for supporting me in everything. And I would like to thank my company, Inmeta, and my managers, Mohammad Yassin and Jon Sandvand, for making me a cloud evangelist.

– Devlin Basilan Duldulao

Table of Contents

3
Connecting to the Database

Section 2: Building the Application

4
Hooking Apollo into React

5

Reusable React Components and React Hooks

6

Authentication with Apollo and React

7

Handling Image Uploads

8

Routing in React

9

Implementing Server-Side Rendering

10

Real-Time Subscriptions

11
Writing Tests for React and Node.js

Section 3: Preparing for Deployment

12
Continuous Deployment with CircleCI and AWS

Index

Other Books You May Enjoy

Preface

The increase in web developers relying on JavaScript to build their frontends and backends has been huge in the last few years. This book covers some of the major technologies from Apollo, Express.js, Node.js, and React. We will go through how to set up React and Apollo to run GraphQL requests against your backend built with Node.js and Sequelize. On top of that, we will introduce testing for the components or functions we have written and automate the deployment on AWS ECS with CircleCI. By the end of the book, you will know how to combine the newest frontend and backend technologies.

Who this book is for

This book is for web developers familiar with React and GraphQL who want to enhance their skills and build full-stack applications using industry standards such as React, Apollo, Node.js, and SQL at scale while learning to solve complex problems with GraphQL.

What this book covers

Chapter 1, *Preparing Your Development Environment*, explains the architecture of an application by going through some core concepts, the complete process, and preparing a working React setup. We will see how React, Apollo Client, and Express.js fit together and cover some good practices when working with React. Further, we will show you how to debug the frontend with React Developer Tools.

Chapter 2, *Setting Up GraphQL with Express.js*, teaches you how to configure your backend by installing Express.js and Apollo via NPM. Express.js will be used for the web server, which handles and passes all GraphQL requests to Apollo.

Chapter 3, *Connecting to the Database*, discusses the opportunities that GraphQL offers when it comes to mutating and querying data. As an example, we will use traditional SQL to build a full application. To simplify the database code, we will use Sequelize, which lets us query our SQL Server with a normal JavaScript object and also allows us to use MySQL, MSSQL, PostgresSQL, or just a SQLite file. We will build models and schemas for users and posts in Apollo and Sequelize.

Chapter 4, Hooking Apollo into React, is where you will learn how to hook Apollo into React and build frontend components to send GraphQL requests. This chapter will explain Apollo-specific configurations.

Chapter 5, Reusable React Components and React Hooks, with the basic concepts and flow of fetching and presenting data clear, will dive deeper into writing more complex React components and sharing data across them.

Chapter 6, Authentication with Apollo and React, will explain the common ways of authenticating a user on the web and in GraphQL. You will be guided through building a complete authentication workflow by using best practices.

Chapter 7, Handling Image Uploads, is the point by which you will have a working authentication and authorization system built on top of Apollo. Moving on, to go beyond normal requests with JSON responses as with GraphQL, we will now upload images via Apollo and save them in separate object storage such as AWS S3.

Chapter 8, Routing in React, is where, to build a complete application for the end user, you will implement some further features, such as a profile page. We will accomplish this by installing React Router v5.

Chapter 9, Implementing Server-Side Rendering, covers server-side rendering. For many applications, this is a must-have. It is important for SEO but can also have positive effects on your end user. This chapter will focus on getting your current application moved to a server-side rendered setup.

Chapter 10, Real-Time Subscriptions, looks at how our application is a great use case for WebSocket and Apollo subscriptions. Many of the applications we use daily have a self-updating notification bar. This chapter will focus on how to build this feature with a more or less experimental GraphQL and Apollo feature called subscriptions.

Chapter 11, Writing Tests for React and Node.js, looks at how a real production-ready application always has an automated testing environment. We will use Mocha, a JavaScript unit testing framework, and Enzyme, a React testing tool, to ensure the quality of our application. This chapter will focus on testing the GraphQL backend and how to properly test React applications with Enzyme.

Chapter 12, Continuous Deployment with CircleCI and AWS, studies deployment. Deploying an application means no more uploading files manually via FTP. Nowadays, you can virtually run your application in the cloud without having a complete server running. For easy deployment of our application, we will use Docker. Before deploying our application, we will quickly cover a basic continuous deployment setup, which will let you deploy all your new code with ease. This chapter will explain how to deploy your applications using Git, Docker, AWS, and CircleCI.

To get the most out of this book

To get started with this book and write functional code, you need to meet some requirements. Regarding the operating system, you can run the complete code and other dependencies on nearly all operating systems out there. The main dependencies of this book are explained one by one within this book.

Software/hardware covered in the book:

- Node.js 14+

- React 17+

- Sequelize 6+

- MySQL 5 or 8

Download the example code files

You can download the example code files for this book from GitHub at `https://github.com/PacktPublishing/Full-Stack-Web-Development-with-GraphQL-and-React-Second-Edition`. If there's an update to the code, it will be updated in the GitHub repository.

We also have other code bundles from our rich catalog of books and videos available at `https://github.com/PacktPublishing/`. Check them out!

Download the color images

We also provide a PDF file that has color images of the screenshots and diagrams used in this book. You can download it here:

`https://static.packt-cdn.com/downloads/9781801077880_ColorImages.pdf`

Conventions used

There are a number of text conventions used throughout this book.

`Code in text`: Indicates code words in text, database table names, folder names, filenames, file extensions, pathnames, dummy URLs, user input, and Twitter handles. Here is an example: "The newest version of Apollo Client comes with the `useQuery` Hook."

A block of code is set as follows:

```
if (loading) return 'Loading...';
if (error) return 'Error! ${error.message}';
```

When we wish to draw your attention to a particular part of a code block, the relevant lines or items are set in bold:

```
mkdir src/client/apollo
touch src/client/apollo/index.js
```

Any command-line input or output is written as follows:

```
mkdir src/client/components
```

Bold: Indicates a new term, an important word, or words that you see onscreen. For instance, words in menus or dialog boxes appear in **bold**. Here is an example: "In the top bar, you will find the **Prettify** button, which tidies your query so that it is more readable."

> **Tips or Important Notes**
> Appear like this.

Get in touch

Feedback from our readers is always welcome.

General feedback: If you have questions about any aspect of this book, email us at customercare@packtpub.com and mention the book title in the subject of your message.

Errata: Although we have taken every care to ensure the accuracy of our content, mistakes do happen. If you have found a mistake in this book, we would be grateful if you would report this to us. Please visit www.packtpub.com/support/errata and fill in the form.

Piracy: If you come across any illegal copies of our works in any form on the internet, we would be grateful if you would provide us with the location address or website name. Please contact us at copyright@packt.com with a link to the material.

If you are interested in becoming an author: If there is a topic that you have expertise in and you are interested in either writing or contributing to a book, please visit authors.packtpub.com.

Share Your Thoughts

Once you've read *Full-Stack Web Development with GraphQL and React Second Edition,* we'd love to hear your thoughts! Scan the QR code below to go straight to the Amazon review page for this book and share your feedback.

https://packt.link/r/1801077886

Your review is important to us and the tech community and will help us make sure we're delivering excellent quality content.

Section 1: Building the Stack

Each journey starts with a first step. Our first step will be to take a look at how to get the basic setup with Node.js, React, MySQL, and GraphQL running. Knowing how to build such a setup yourself and how the different technologies work together is very important to understand more advanced topics later in the book.

In this section, there are the following chapters:

- *Chapter 1, Preparing Your Development Environment*
- *Chapter 2, Setting Up GraphQL with Express.js*
- *Chapter 3, Connecting to the Database*

1
Preparing Your Development Environment

The application we are going to build in this book will be a simplified version of Facebook, called **Graphbook**. We will allow our users to sign up and log in to read and write posts and chat with friends, similar to what we can do on common social networks.

When developing an application, being well-prepared is always a requirement. However, before we start, we need to put our stack together. In this chapter, we will explore whether our techniques work well with our development process, what we need before getting started, and which tools can help us when building software.

This chapter explains the architecture for our application by going through the core concepts, the complete process, and preparing for a working React setup.

This chapter covers the following topics:

- Architecture and technology
- Thinking critically about how to architect a stack
- Building the React and GraphQL stack

- Installing and configuring Node.js
- Setting up a React development environment with webpack, Babel, and other requirements
- Debugging React applications using Chrome DevTools and React Developer Tools

Technical requirements

The source code for this chapter is available in the following GitHub repository: `https://github.com/PacktPublishing/Full-Stack-Web-Development-with-GraphQL-and-React-Second-Edition/tree/main/Chapter01`.

Understanding the application architecture

Since its initial release in 2015, GraphQL has become the new alternative to the standard SOAP and REST APIs. GraphQL is a specification, like SOAP and REST, that you can follow to structure your application and data flow. It is so innovative because it allows you to query specific fields of entities, such as users and posts. This functionality makes it very good for targeting multiple platforms at the same time. Mobile apps may not need all the data that's displayed inside the browser on a desktop computer. The query you send consists of a JSON-like object that defines which information your platform requires. For example, a query for a `post` may look like this:

```
post {
  id
  text
  user {
    user_id
    name
  }
}
```

GraphQL resolves the correct entities and data, as specified in your query object. Every field in GraphQL represents a function that resolves to a value. Those functions are called **Resolver functions**. The return value could be just the corresponding database value, such as the name of a user, or it could be a date, which is formatted by your server before returning it.

GraphQL is completely database agnostic and can be implemented in any programming language. To skip the step of implementing a GraphQL library, we are going to use Apollo, which is a GraphQL server for the Node.js ecosystem. Thanks to the team behind Apollo, this is very modular. Apollo works with many of the common Node.js frameworks, such as Hapi, Koa, and Express.js.

We are going to use Express.js as our basis because it is used on a wide scale in the Node.js and GraphQL communities. GraphQL can be used with multiple database systems and distributed systems to offer a straightforward API over all your services. It allows developers to unify existing systems and handle data fetching for client applications.

How you combine your databases, external systems, and other services into one server backend is up to you. In this book, we are going to use a MySQL server via Sequelize as our data storage. SQL is the most well-known and commonly used database query language, and with Sequelize, we have a modern client library for our Node.js server to connect with our SQL server.

HTTP is the standard protocol for accessing a GraphQL API. It also applies to Apollo Servers. However, GraphQL is not fixed to one network protocol. Everything we have mentioned so far is everything important for the backend.

When we get to the frontend of our **Graphbook** application, we are mainly going to use React. React is a JavaScript UI framework that was released by Facebook that has introduced many techniques that are now commonly used for building interfaces on the web, as well as on native environments.

Using React comes with a bunch of significant advantages. When building a React application, you always split your code into many components, targeting their efficiency and ability to be reused. Of course, you can do this without using React, but it makes it very easy. Furthermore, React teaches you how to update application states, as well as the UI, reactively. You never update the UI and then the data separately.

React makes rerendering very efficient by using a virtual DOM, which compares the virtual and actual DOM and updates it accordingly. Only when there is a difference between the virtual and real DOM does React apply these changes. This logic stops the browser from recalculating the layout, **Cascading Style Sheets** (**CSS**), and other computations that negatively impact the overall performance of your application.

Throughout this book, we are going to use the Apollo client library. It naturally integrates with React and our Apollo Server.

If we put all this together, the result is the main stack consisting of Node.js, Express.js, Apollo, SQL, Sequelize, and React.

The basic setup

The basic setup for making an application work is the logical request flow, which looks as follows:

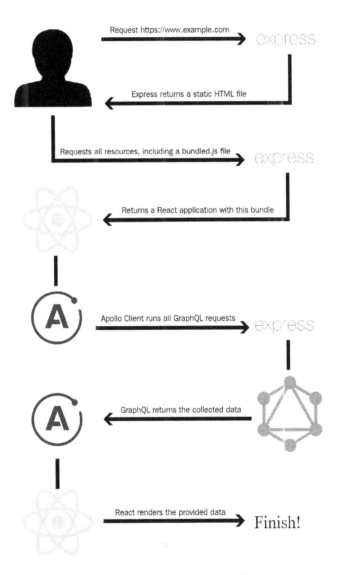

Figure 1.1 – Logical request workflow

Here is how the logical request flow works:

1. The client requests our site.
2. The Express.js server handles these requests and serves a static HTML file.

3. The client downloads all the necessary files, according to this HTML file. The files also include a bundled JavaScript file.

4. This bundled JavaScript file is our React application. After executing all the JavaScript code from this file, all the required Ajax alias GraphQL requests are made to our Apollo Server.

5. Express.js receives the requests and passes them to our Apollo endpoint.

6. Apollo queries all the requested data from all the available systems, such as our SQL server or third-party services, merges the data, and sends it back as JSON.

7. React can render the JSON data as HTML.

This workflow is the basic setup for making an application work. In some cases, it makes sense to offer server-side rendering for our client. The server would need to render and send all `XMLHttpRequests` itself before returning the HTML to the client. The user will save doing one or more round trips if the server sends the requests on the initial load. We will focus on this topic later in this book, but that is the application architecture in a nutshell. With that in mind, let's get hands-on and set up our development environment.

Installing and configuring Node.js

The first step of preparing our project is to install Node.js. There are two ways to do this:

- One option is to install **Node Version Manager** (**NVM**). The benefit of using NVM is that you can easily run multiple versions of Node.js side by side, which handles the installation process for you on nearly all UNIX-based systems, such as Linux and macOS. Within this book, we do not need the option to switch between different versions of Node.js.

- The other option is to install Node.js via the package manager of your distribution if you are using Linux. The official PKG file is for Mac, while the MSI file is for Windows. We are going to use the regular Linux package manager for this book as it is the easiest method.

> **Note**
> You can find the **Downloads** section of Node.js at the following link: `https://nodejs.org/en/download/`.

We are going to be using the second option here. It covers the regular server configurations and is easy to understand. I will keep this as short as possible and skip all the other options, such as Chocolatey for Windows and Brew for Mac, which are very specialized for those specific operating systems.

I assume that you are using a Debian-based system for ease of use with this book. It has got the normal APT package manager and repositories for easily installing Node.js and MySQL. If you are not using a Debian-based system, you can look up the matching commands to install Node.js at `https://nodejs.org/en/download/package-manager/`.

Our project is going to be new so that we can use Node.js 14, which is the current LTS version:

1. First, let's add the correct repository for our package manager by running the following command:

    ```
    curl -fsSL https://deb.nodesource.com/setup_14.x | sudo
    bash -
    ```

2. Next, we must install Node.js and the build tools for native modules using the following command:

    ```
    apt-get install -y nodejs build-essential
    ```

3. Finally, let's open a Terminal and verify that the installation was successful:

    ```
    node --version
    ```

> Note
> Installing Node.js via the package manager will automatically install npm.

Great! You are now set to run server-side JavaScript with Node.js and install Node.js modules for your projects with npm.

All the dependencies that our project relies on are available at `https://npmjs.com` and can be installed with npm or Yarn. We will rely on npm as it is more widely used than Yarn. So, let's continue and start using npm to set up our project and its dependencies.

Setting up React

The development environment for our project is ready. In this section, we are going to install and configure React, which is the primary aspect of this chapter. Let's start by creating a new directory for our project:

```
mkdir ~/graphbook
cd ~/graphbook
```

Our project will use Node.js and many npm packages. We will create a `package.json` file to install and manage all the dependencies for our project. It stores information about the project, such as the version number, name, dependencies, and much more.

Just run `npm init` to create an empty `package.json` file:

```
npm init
```

npm will ask some questions, such as asking for the package name, which is, in fact, the project name. Enter `Graphbook` to insert the name of your application in the generated `package.json` file.

I prefer to start with version number 0.0.1 since the default version number that npm offers with 1.0.0 represents the first stable release for me. However, it is your choice regarding which version you use here.

You can skip all the other questions by pressing the *Enter* key to save the default values of npm. Most of them are not relevant because they just provide information such as a description or the link to the repository. We are going to fill in the other fields, such as the scripts, while working through this book. You can see an example of the command line in the following screenshot:

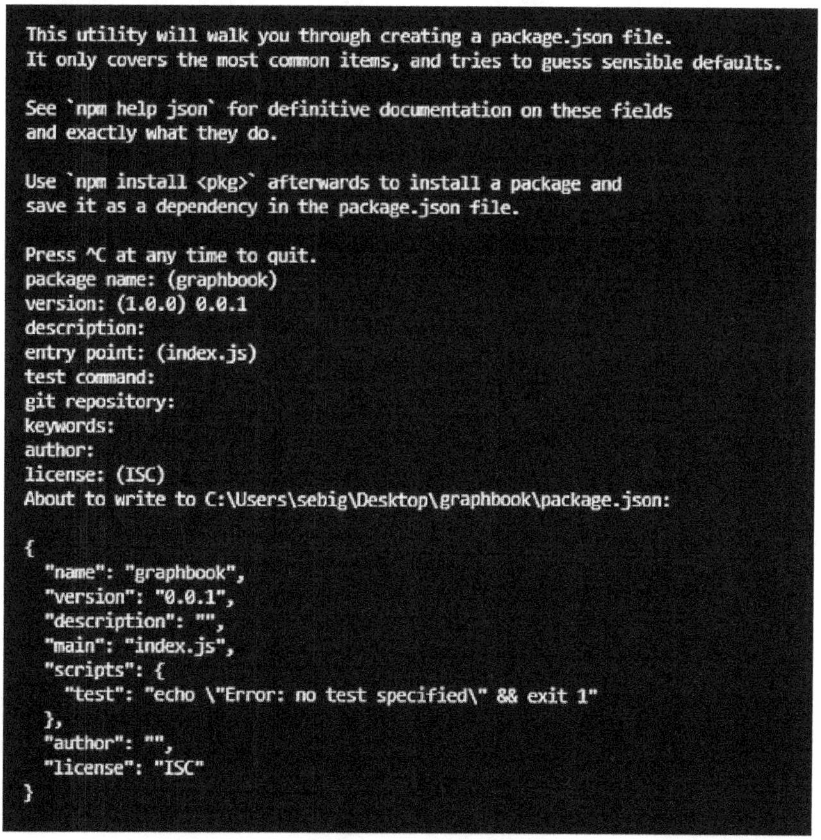

```
This utility will walk you through creating a package.json file.
It only covers the most common items, and tries to guess sensible defaults.

See `npm help json` for definitive documentation on these fields
and exactly what they do.

Use `npm install <pkg>` afterwards to install a package and
save it as a dependency in the package.json file.

Press ^C at any time to quit.
package name: (graphbook)
version: (1.0.0) 0.0.1
description:
entry point: (index.js)
test command:
git repository:
keywords:
author:
license: (ISC)
About to write to C:\Users\sebig\Desktop\graphbook\package.json:

{
  "name": "graphbook",
  "version": "0.0.1",
  "description": "",
  "main": "index.js",
  "scripts": {
    "test": "echo \"Error: no test specified\" && exit 1"
  },
  "author": "",
  "license": "ISC"
}
```

Figure 1.2 – npm project setup

The first and most crucial dependency for this book is React. Use npm to add React to our project:

```
npm install --save react react-dom
```

This command installs two npm packages from `https://npmjs.com` into our project folder under `node_modules`.

npm automatically edited our `package.json` file since we provided the `--save` option and added those packages with the latest available version numbers.

You might be wondering why we installed two packages, even though we only needed React. The `react` package only provides React-specific methods. All React Hooks, such as `componentDidMount`, `useState`, and even React's component class, come from this package. You need this package to write any React application.

In most cases, you won't even notice that you have used `react-dom`. This package offers all the functions to connect the actual DOM of the browser to your React application. Usually, you use `ReactDOM.render` to render your application at a specific point in your HTML and only once in your code. We will cover how to render React later in this book.

There is also a function called `ReactDOM.findDOMNode`, which gives you direct access to a DOMNode, but I strongly discourage using this since any changes on DOMNodes are not available in React itself. I have never needed to use this function, so try to avoid it if possible. Now, that our npm project has been set up and the two main dependencies have been installed, we need to prepare an environment that bundles all the JavaScript files we are going to write. We will focus on this in the next section.

Preparing and configuring webpack

Our browser requests an `index.html` file when accessing our application. It specifies all the files that are required to run our application. We need to create this `index.html` file, which we serve as the entry point of our application:

1. Create a separate directory for our `index.html` file:

    ```
    mkdir public
    cd public
    touch index.html
    ```

2. Then, save this inside `index.html`:

```html
<!DOCTYPE html>
<html lang="en">
  <head>
    <meta charset="UTF-8">
    <meta name="viewport" content="width=device-width,
      initial-scale=1.0">
    <meta http-equiv="X-UA-Compatible" content="ie=edge">
    <title>Graphbook</title>
  </head>
  <body>
    <div id="root"></div>
  </body>
</html>
```

As you can see, no JavaScript is loaded here. There is only one `div` tag with the `root` ID. This `div` tag is the DOMNode that our application will be rendered in by `ReactDOM`.

So, how do we get React up and running with this `index.html` file?

To accomplish this, we need to use a web application bundler, which will prepare and bundle all our application assets. All the required JavaScript files and `node_modules` are bundled and minified; SASS and SCSS preprocessors are transpiled to CSS, as well as being merged and minified.

A few application bundler packages are available, including webpack, Parcel, and Gulp. For our use case, we will use webpack. It is the most common module bundler and has a large community surrounding it. To bundle our JavaScript code, we need to install webpack and all its dependencies, as follows:

```
npm install --save-dev @babel/core babel-loader @babel/preset-
env @babel/preset-react clean-webpack-plugin css-loader file-
loader html-webpack-plugin style-loader url-loader webpack
webpack-cli webpack-dev-server
```

This command adds all the development tools to `devDependencies` in the `package.json` file. We will need these to bundle our application. They are only installed in a development environment and are skipped in production.

If you are not already aware, setting up webpack can be a bit of a hassle. Many options can interfere with each other and lead to problems when you're bundling your application. Now, let's create a `webpack.client.config.js` file in the root folder of your project.

Enter the following code:

```
const path = require('path');
const HtmlWebpackPlugin = require('html-webpack-plugin');
const { CleanWebpackPlugin } = require('clean-webpack-plugin');
const buildDirectory = 'dist';
const outputDirectory = buildDirectory + '/client';
module.exports = {
  mode: 'development',
  entry: './src/client/index.js',
  output: {
    path: path.join(__dirname, outputDirectory),
    filename: 'bundle.js'
  },
  module: {
    rules: [
      {
        test: /\.js$/,
        exclude: /node_modules/,
        use: {
          loader: 'babel-loader'
        }
      },
      {
        test: /\.css$/,
        use: ['style-loader', 'css-loader']
      }
    ]
  },
  devServer: {
    port: 3000,
    open: true
  },
```

```
  plugins: [
    new CleanWebpackPlugin({
      cleanOnceBeforeBuildPatterns: [path.join(__dirname,
      buildDirectory)]
    }),
    new HtmlWebpackPlugin({
      template: './public/index.html'
    })
  ]
};
```

The webpack configuration file is just a regular JavaScript file where you can require node_modules and custom JavaScript files. This is the same as everywhere else inside Node.js. Let's quickly go through all of the main properties of this configuration. Understanding these will make future custom webpack configs much easier. All the important points are explained here:

- HtmlWebpackPlugin: This automatically generates an HTML file that includes all the webpack bundles. We pass our previously created index.html as a template.

- CleanWebpackPlugin: This empties all of the provided directories to clean the old build files. The cleanOnceBeforeBuildPatterns property specifies an array of folders that are cleaned before the build process is started.

- The entry field tells webpack where the starting point of our application is. This file needs to be created by us.

- The output object specifies how our bundle is called and where it should be saved. For us, this is dist/client/bundle.js.

- Inside module.rules, we match our file extensions with the correct loaders. All the JavaScript files (except those located in node_modules) are transpiled by Babel as specified by babel-loader so that we can use ES6 features inside our code. Our CSS gets processed by style-loader and css-loader. There are many more loaders for JavaScript, CSS, and other file extensions available.

- The devServer feature of webpack enables us to run the React code directly. This includes hot reloading code in the browser without rerunning a build or refreshing the browser tab.

> **Note**
> If you need a more detailed overview of the webpack configuration, have a look at the official documentation: `https://webpack.js.org/concepts/`.

With this in mind, let's move on. We are missing the `src/client/index.js` file from our webpack configuration, so let's create it, as follows:

```
mkdir -p src/client
cd src/client
touch index.js
```

You can leave this file empty for the moment. It can be bundled by webpack without content inside. We are going to change it later in this chapter.

To spin up our development webpack server, we will add a command to `package.json` that we can run using npm.

Add the following line to the `scripts` object inside `package.json`:

```
"client": "webpack serve --devtool inline-source-map --hot
--config webpack.client.config.js"
```

Now, execute npm `run client` in your console and watch how a new browser window opens. We are running `webpack serve` with the newly created configuration file.

Sure, the browser is still empty, but if you inspect the HTML with Chrome DevTools, you will see that we have already got a `bundle.js` file and that our `index.html` file was taken as a template.

With that, we've learned how to include our empty `index.js` file with the bundle and serve it to the browser. Next, we'll render our first React component inside our template `index.html` file.

Rendering your first React component

There are many best practices for React. The central philosophy behind it is to split our code into separate components where possible. We are going to cover this approach in more detail in *Chapter 5, Reusable React Components and React Hooks*.

Our `index.js` file is the main starting point of our frontend code, and this is how it should stay. Do not include any business logic in this file. Instead, keep it as clean and slim as possible.

The `index.js` file should include the following code:

```
import React from 'react';
import ReactDOM from 'react-dom';
import App from './App';


ReactDOM.render(<App/>, document.getElementById('root'));
```

The release of *ECMAScript 2015* introduced the `import` feature. We can use it to load our npm packages – `react` and `react-dom` – and our first custom React component, which we must write now.

Of course, we need to cover the traditional `Hello World` program.

Create the `App.js` file next to your `index.js` file and ensure it contains the following content:

```
import React from 'react';


const App = () => {
    return (
        <div>Hello World!</div>
    )
}
export default App
```

Here, we exported a single stateless function called `App` that is then imported by the `index.js` file. As we explained previously, we are now actively using `ReactDOM.render` in our `index.js` file.

The first parameter of `ReactDOM.render` is the component or function that we want to render, which is the exported function displaying the **Hello World!** message. The second parameter is the browser's `DOMNode`, where it should render. We receive `DOMNode` with a plain `document.getElementById` JavaScript.

We defined our root element when we created the `index.html` file. After saving the `App.js` file, webpack will try to build everything again. However, it should not be able to do that. webpack will encounter a problem when bundling our `index.js` file because of the `<App />` tag syntax we are using in the `ReactDOM.render` method. It was not transpiled to a normal JavaScript function.

We configured webpack to load Babel for our JavaScript file but did not tell Babel what to transpile and what not to transpile.

Let's create a `.babelrc` file in the root folder that contains the following content:

```
{
  "presets": ["@babel/env","@babel/react"]
}
```

> **Note**
> You may have to restart the server because the `.babelrc` file is not reloaded when changes are made to the file. After a few moments, you should see the standard **Hello World!** message in your browser.

Here, we told Babel to use `@babel/preset-env` and `@babel/preset-react`, which we installed together with webpack. These presets allow Babel to transform specific syntax, such as JSX. We can use those presets to create normal JavaScript that all browsers can understand and that webpack can bundle.

Rendering arrays from React state

`Hello World!` is a must for every good programming book, but this is not what we are aiming for when we use React.

A social network such as Facebook or Graphbook, which we are writing now, needs a newsfeed and an input to post news. Let's implement this.

Since this is the first chapter of this book, we will do this inside `App.js`.

We should work with some fake data here since we have not set up our GraphQL API yet. We can replace this with real data later.

Define a new variable above the `default` exported `App` function, like this:

```
const initialPosts = [
    {
        id: 2,
        text: 'Lorem ipsum',
        user: {
        avatar: '/uploads/avatar1.png',
        username: 'Test User'
        }
    },
    {
```

```
        id: 1,
        text: 'Lorem ipsum',
        user: {
        avatar: '/uploads/avatar2.png',
        username: 'Test User 2'
        }
    }
];
```

We are going to render these two fake posts in React. To prepare this, change the first line of the App.js file to the following:

```
import React, { useState } from 'react';
```

This ensures that the useState function of React is imported and accessible by our stateless function.

Replace the current content of your App function with the following code:

```
const [posts, setPosts] = useState(initialPosts);

return (
  <div className="container">
    <div className="feed">
      { initialPosts.map((post, i) =>
        <div key={post.id} className="post">
          <div className="header">
            <img src={post.user.avatar} />
            <h2>{post.user.username}</h2>
          </div>
          <p className="content">
            {post.text}
          </p>
        </div>
      )}
    </div>
  </div>
)
```

Here, we initiated a `posts` array inside the function via the React `useState` function. It allows us to have a state without writing a real React class; instead, it only relies on raw functions. The `useState` function expects one parameter, which is the initial value of the state variable. In this case, this is the constant `initialPosts` array. This returns the `posts` state variable and a `setPosts` function, which you can use to update the local state.

Then, we iterated over the `posts` array with the `map` function, which again executes the inner callback function, passing each array item as a parameter one by one. The second parameter is just called `i` and represents the index of the array element we are processing. Everything that's returned from the `map` function is then rendered by React.

We merely returned HTML by putting each post's data in ES6 curly braces. These curly braces tell React to interpret and evaluate the code inside them as JavaScript.

As shown in the preceding code, we are relying on the posts that were returned by the `useState` function. This data flow is very convenient because we can update the state at any point in our application and the posts will rerender. The important thing is that this will only work by using the `setPosts` function and passing the updated array to it. In this case, React notices the change of state and rerenders the function.

The preceding method is much cleaner, and I recommend this for readability purposes. When saving, you should be able to see rendered posts. They should look like this:

Test User
Lorem ipsum

Test User 2
Lorem ipsum

Figure 1.3 – Unstyled demo posts

The images I am using here are freely available. You can use any other material that you have got if the path matches the string from the `posts` array. You can find those images in the official GitHub repository for this book.

CSS with webpack

The posts in the preceding figure have not been designed yet. I have already added CSS classes to the HTML our component returns.

Instead of using CSS to make our posts look better, another method is to use CSS-in-JS using packages such as styled components, which is a React package. Other alternatives include Glamorous and Radium. There are numerous reasons for and against using such libraries. With those other tools, you are not able to use SASS, SCSS, or LESS effectively. I need to work with other people, such as screen and graphics designers, who can provide and use CSS, but do not program styled components. There is always a prototype or existing CSS that can be used, so why should I spend time translating this into styled components CSS when I could just continue with standard CSS?

There is no right or wrong option here; you are free to implement the styling in any way you like. However, in this book, we will keep using good old CSS.

What we have already done in our `webpack.client.config.js` file is specify a CSS rule, as shown in the following code snippet:

```
{
    test: /\.css$/,
    use: ['style-loader', 'css-loader'],
},
```

`style-loader` injects your bundled CSS right into the DOM. `css-loader` will resolve all `import` or `url` occurrences in your CSS code.

Create a `style.css` file in `/assets/css` and fill in the following:

```
body {
    background-color: #f6f7f9;
    margin: 0;
    font-family: 'Courier New', Courier, monospace
}
p {
    margin-bottom: 0;
}
```

```css
.container {
  max-width: 500px;
  margin: 70px auto 0 auto;
}
.feed {
  background-color: #bbb;
  padding: 3px;
  margin-top: 20px;
}
.post {
  background-color: #fff;
  margin: 5px;
}
.post .header {
  height: 60px;
}
.post .header > * {
  display: inline-block;
  vertical-align: middle;
}
.post .header img {
  width: 50px;
  height: 50px;
  margin: 5px;
}
.post .header h2 {
  color: #333;
  font-size: 24px;
  margin: 0 0 0 5px;
}
.post p.content {
  margin: 5px;
  padding: 5px;
  min-height: 50px;
}
```

Refreshing your browser will leave you with the same old HTML you had previously.

This problem occurs because webpack is a module bundler and does not know anything about CSS; it only knows JavaScript. We must import the CSS file somewhere in our code.

Instead of using index.html and adding a head tag, we can use webpack and our CSS rule to load it right into App.js. This solution is very convenient since all the required CSS throughout our application gets minified and bundled. Webpack automates this process.

In your App.js file, add the following behind the React import statement:

```
import '../../assets/css/style.css';
```

webpack magically rebuilds our bundle and refreshes our browser tab.

With that, you have successfully rendered fake data via React and styled it with bundled CSS from webpack. It should look something like this:

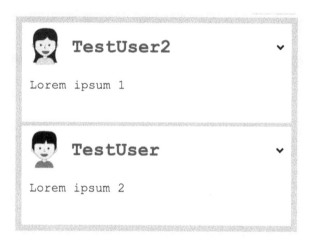

Figure 1.4 – Styled demo posts

The output looks very good already.

Event handling and state updates with React

For this project, it would be great to have a simple textarea where we can click a button and then add a new post to the static posts array we wrote in the App function.

Add the following code above the div tag that contains the feed class:

```
<div className="postForm">
  <form onSubmit={handleSubmit}>
    <textarea value={postContent} onChange={ (e) =>
      setPostContent(e.target.value)}
      placeholder="Write your custom post!"/>
    <input type="submit" value="Submit" />
  </form>
</div>
```

You can use forms in React without any problems. React can intercept the submit event of requests by giving the form an onSubmit property, which will be a function to handle the logic.

We are passing the postContent variable to the value property of textarea to get what is called a **controlled input**.

Create an empty string variable to save the textarea value by using the useState function from React:

```
const [postContent, setPostContent] = useState('');
```

The postContent variable is already being used for our new textarea since we specified it in the value property. Furthermore, we directly implemented the setPostContent function in our post form. This is used for the onChange property or any event that is called whenever you type inside textarea. The setPostContent function receives the e.target.value variable, which is the DOM accessor for the value of textarea, which is then stored in the state of the React function.

Look at your browser again. The form is there, but it is not pretty, so add the following CSS:

```
form {
  padding-bottom: 20px;
}
form textarea {
  width: calc(100% - 20px);
  padding: 10px;
  border-color: #bbb;
}
form [type=submit] {
```

```
  border: none;
  background-color: #6ca6fd;
  color: #fff;
  padding: 10px;
  border-radius: 5px;
  font-size: 14px;
  float: right;
}
```

The last step is to implement the `handleSubmit` function for our form. Add it straight after the state variables and the `return` statement:

```
const handleSubmit = (event) => {
    event.preventDefault();
    const newPost = {
        id: posts.length + 1,
        text: postContent,
        user: {
            avatar: '/uploads/avatar1.png',
            username: 'Fake User'
        }
    };
    setPosts([newPost, ...posts]);
    setPostContent('');
};
```

The preceding code looks more complicated than it is, but I am going to explain it quickly.

We needed to run `event.preventDefault` to stop our browser from actually trying to submit the form and reload the page. Most people that come from jQuery or other JavaScript frameworks will know this.

Next, we saved our new post in the `newPost` variable, which we want to add to our feed.

We faked some data here to simulate a real-world application. For our test case, the new post ID is the number of posts in our state variable plus one. React wants us to give every child in the ReactDOM a unique ID. By counting the number of posts with `posts.length`, we simulate the behavior of a real backend giving us unique IDs for our posts.

The text for our new post comes from the `postContent` state variable.

Furthermore, we do not have a user system right now that our GraphQL server can use to give us the newest posts, including the matching users and their avatars. We can simulate this by having a static user object for all the new posts we create.

Finally, we updated the state again. We did this by using the `setPosts` function and passing a merged array consisting of the new posts and the current `posts` array with a destructuring assignment. After that, we cleared `textarea` by passing an empty string to the `setPostContent` function.

Now, go ahead and play with your working React form. Do not forget that all the posts you create do not persist since they are only held in the local memory of the browser and are not saved to a database. Consequently, refreshing deletes your posts.

Controlling document heads with React Helmet

When developing a web application, you must control your document heads. You might want to change the title or description, based on the content you are presenting.

React Helmet is a great package that offers this on the fly, including overriding multiple headers and server-side rendering. Let's see how we can do this:

1. Install React Helmet with the following command:

    ```
    npm install --save react-helmet
    ```

    You can add all standard HTML headers with React Helmet.

    I recommend keeping standard `head` tags inside your template. This has the advantage that, before React has been rendered, there is always the default document head. For our case, you can directly apply a title and description in `App.js`.

2. Import `react-helmet` at the top of the file:

    ```
    import { Helmet } from 'react-helmet';
    ```

3. Add `Helmet` directly above `postForm` div:

    ```
    <Helmet>
      <title>Graphbook - Feed</title>
      <meta name="description" content="Newsfeed of all
        your friends on Graphbook" />
    </Helmet>
    ```

If you reload the browser and watch the title on the tab bar of your browser carefully, you will see that it changes from Graphbook to Graphbook - Feed. This behavior happens because we already defined a title inside index.html. When React finishes rendering, the new document head is applied.

Production build with webpack

The last step for our React setup is to have a production build. Until now, we were only using webpack-dev-server, but this naturally includes an unoptimized development build. Furthermore, webpack automatically spawns a web server. In the next chapter, we will introduce Express.js as our web server so that we don't need webpack to start it.

A production bundle does merge all JavaScript files, but it also merges all the CSS files into two separate files. Those can be used directly in the browser. To bundle CSS files, we will rely on another webpack plugin, called MiniCss:

```
npm install --save-dev mini-css-extract-plugin
```

We do not want to change the current webpack.client.config.js file because it is made for development work. Add the following command to the scripts object of your package.json:

```
"client:build": "webpack --config webpack.client.build.config.
js"
```

This command runs webpack using an individual production webpack config file. Let's create this one. First, clone the original webpack.client.config.js file and rename it webpack.client.build.config.js.

Change the following things in the new file:

1. mode needs to be production, not development.

2. Require the MiniCss plugin:

    ```
    const MiniCssExtractPlugin = require('mini-css-extract-
    plugin');
    ```

3. Replace the current CSS rule:

    ```
    {
        test: /\.css$/,
        use: [{ loader: MiniCssExtractPlugin.loader,
          options: {
    ```

```
        publicPath: '../'
      }
    }, 'css-loader'],
  },
```

We are no longer using `style-loader`; instead, we're using the `MiniCss` plugin. The plugin goes through the complete CSS code, merges it into a separate file, and removes the `import` statements from the `bundle.js` file, which we generate in parallel.

4. Lastly, add the plugin to the plugins at the bottom of the configuration file:

```
new MiniCssExtractPlugin({
  filename: 'bundle.css',
})
```

5. Remove the entire `devServer` property.

When you run the new configuration, it won't spawn a server or browser window; it will only create a production JavaScript and CSS bundle and will require them in our `index.html` file. According to our `webpack.client.build.config.js` file, those three files are going to be saved to the `dist/client` folder.

You can run this command by executing `npm run client:build`.

If you look in the `dist/client` folder, you will see three files. You can open the `index.html` file in your browser. Sadly, the images are broken because the image URLs are not correct anymore. We must accept this for the moment because it will be automatically fixed when we have a working backend.

With that, we have finished the basic setup for React.

Useful development tools

When you're working with React, you want to know why your application rendered the way that it did. You need to know which properties your components received and how their current state looks. Since this is not displayed in the DOM or anywhere else in Chrome DevTools, you need a separate plugin.

Fortunately, Facebook has got you covered. Visit `https://chrome.google.com/webstore/detail/react-developer-tools/fmkadmapgofadopljbjfkapdkoienihi` and install React Developer Tools. This plugin allows you to inspect React applications and components. When you open Chrome DevTools again, you will see that there are two new tabs at the end of the row – one called **Components** and another called **Profiler**:

Figure 1.5 – React developer tools

You will only be able to see those tabs if you are running a React application in development mode. If a React application is running or bundled in production, those extensions won't work.

> **Note**
> If you are unable to see this tab, you may need to restart Chrome completely. You can also find React Developer Tools for Firefox.

The first tab allows you to view, search, and edit all the components of your ReactDOM.

The left-hand side panel looks much like the regular DOM tree (Elements) in Chrome DevTools, but instead of showing HTML markup, you will see all the components you used inside a tree. ReactDOM rendered this tree into real HTML, as follows:

Figure 1.6 – React component tree

The first component in the current version of Graphbook should be `<App />`.

By clicking a component, your right-hand side panel will show its properties, state, and context. You can try this with the App component, which is the only real React component:

Figure 1.7 – React component state

The App function is the first component of our application. This is the reason why it received no props. Children can receive properties from their parents; with no parent, there are no props.

Now, test the App function and play around with the state. You will see that changing it rerenders your ReactDOM and updates the HTML. You can edit the postContent variable, which inserts the new text inside textarea. As you will see, all the events are thrown, and your handler runs. Updating the state always triggers a rerender, so try to update the state as little as possible to use as few computing resources as possible.

Summary

In this chapter, we set up a working React environment. This is a good starting point for our frontend as we can write and build static web pages with this setup.

The next chapter will primarily focus on our setup for the backend. We will configure Express.js to accept our first requests and pass all GraphQL queries to Apollo. Furthermore, you will learn how to use Postman to test your API.

2
Setting Up GraphQL with Express.js

The basic setup and prototype for our frontend are now complete. Now, we need to get our GraphQL server running to begin implementing the backend. Apollo and Express.js are going to be used to build the foundation of our backend.

This chapter will explain the installation process for Express.js, as well as the configuration for our GraphQL endpoint. We will quickly go through all the essential features of Express.js and the debugging tools for our backend.

This chapter covers the following topics:

- Express.js installation and explanation
- Routing in Express.js
- Middleware in Express.js
- Binding Apollo Server to a GraphQL endpoint
- Sending our first GraphQL requests
- Backend debugging and logging

Technical requirements

The source code for this chapter is available in the following GitHub repository:

```
https://github.com/PacktPublishing/Full-Stack-Web-Development-
with-GraphQL-and-React-Second-Edition/tree/main/Chapter02
```

Getting started with Node.js and Express.js

One of the primary goals of this book is to set up a GraphQL API, which is then consumed by our React frontend. To accept network requests – especially GraphQL requests – we are going to set up a Node.js web server.

The most significant competitors in the Node.js web server area are Express.js, Koa, and Hapi. In this book, we are going to use Express.js. Most tutorials and articles about Apollo rely on it.

Express.js is also the most used Node.js web server out there and describes itself as a Node.js web framework, offering all the main features needed to build web applications.

Installing Express.js is easy. We can use npm in the same way as we did in the previous chapter:

```
npm install --save express
```

This command adds the latest version of Express to package.json.

In the previous chapter, we created all the JavaScript files directly in the src/client folder. Now, let's create a separate folder for our server-side code. This separation gives us a tidy directory structure. We can create this folder with the following command:

```
mkdir src/server
```

Now, we can continue configuring Express.js.

Setting up Express.js

As always, we need a root file that's loaded with all the main components to combine them into a real application.

Create an `index.js` file in the `server` folder. This file is the starting point for the backend. Here's how we do this:

1. First, we must import `express` from `node_modules`, which we just installed:

    ```
    import express from 'express';
    ```

    We can use `import` here since our backend gets transpiled by Babel. We are also going to set up webpack for the server-side code later in *Chapter 9, Implementing Server-Side Rendering*.

2. Next, we must initialize the server with the `express` command. The result is stored in the `app` variable. Everything our backend does is executed through this object:

    ```
    const app = express();
    ```

3. Then, we must specify the routes that accept requests. For this straightforward introduction, we accept all HTTP `GET` requests that match any path by using the `app.get` method. Other HTTP methods can be caught with `app.post` and `app.put`:

    ```
    app.get('*', (req, res) => res.send('Hello World!'));
    app.listen(8000, () => console.log('Listening on port
    8000!'));
    ```

To match all the paths, you can use an asterisk, which generally stands for `any` in the programming field, as we did in the preceding `app.get` line.

The first parameter for all the `app.METHOD` functions is the path to match. From here, you can provide an unlimited list of callback functions, which are executed one by one. We are going to look at this feature later in the *Routing with Express.js* section.

A callback always receives the client request as the first parameter and the response as the second parameter, which the server is going to send. Our first callback is going to use the `send` response method.

The `send` function merely sends the HTTP response. It sets the HTTP body as specified. So, in our case, the body shows `Hello World!`, while the `send` function takes care of all the necessary standard HTTP headers, such as `Content-Length`.

The last step is to tell Express.js which port the server should listen for requests on. In our code, we are using `8000` as the first parameter of `app.listen`. You can replace `8000` with any port or URL you want to listen on. The callback is executed when the HTTP server is bound to that port and requests can be accepted.

This is the easiest setup we can have for Express.js.

Running Express.js in development

To launch our server, we have to add a new script to our `package.json` file.

Let's add the following line to the `scripts` property of the `package.json` file:

```
"server": "nodemon --exec babel-node --watch src/server src/
server/index.js"
```

As you can see, we are using a command called `nodemon`. We need to install it first:

```
npm install --save nodemon
```

`nodemon` is an excellent tool for running a Node.js application. It can restart your server when the source changes.

For example, to get the preceding command working, follow these steps:

1. First, we must install the `@babel/node` package since we are transpiling the backend code with Babel using the `--exec babel-node` option. This allows us to use the `import` statement:

   ```
   npm install --save-dev @babel/node
   ```

2. Providing the `--watch` option to `nodemon` when following a path or file will permanently track the changes on that file or folder and reload the server to represent the latest state of your application. The last parameter refers to the actual file being the starting execution point for the backend.

3. Start the server:

   ```
   npm run server
   ```

Now, when you go to your browser and enter `http://localhost:8000`, you will see the text **Hello World!** from our Express.js callback function.

Chapter 3, *Connecting to the Database*, covers how Express.js routing works in detail.

Routing in Express.js

Understanding routing is essential to extending our backend code. In this section, we are going to play through some simple routing examples.

In general, routing handles how and where an application responds to specific endpoints and methods.

In Express.js, one path can respond to different HTTP methods and can have multiple handler functions. These handler functions are executed one by one in the order they were specified in the code. A path can be a simple string, but also a complex regular expression or pattern.

When you're using multiple handler functions – either provided as an array or as multiple parameters – be sure to pass `next` to every callback function. When you call `next`, you hand over the execution from one callback function to the next function in the row. These functions can also be middleware. We'll cover this in the next section.

Here is a simple example. Replace this with the current `app.get` line:

```
app.get('/', function (req, res, next) {
  console.log('first function');
  next();
}, function (req, res) {
  console.log('second function');
  res.send('Hello World!');
});
```

When you refresh your browser, look at the server logs in the Terminal; you will see that both `first function` and `second function` are printed. If you remove the execution of `next` and try to reload the browser tab, the request will time out and only `first function` will be printed. This problem occurs because neither `res.send` nor `res.end`, nor any alternative, is called. The second handler function is never executed when `next` is not run.

As we mentioned previously, the **Hello World!** message is nice but not the best we can get. In development, it is completely okay for us to run two separate servers – one for the frontend and one for the backend.

Serving our production build

We can serve our production build of the frontend through Express.js. This approach is not great for development purposes but is useful for testing the build process and seeing how our live application will act.

Again, replace the previous routing example with the following code:

```
import path from 'path';

const root = path.join(__dirname, '../../');
```

```
app.use('/', express.static(path.join(root, 'dist/client')));
app.use('/uploads', express.static(path.join(root,
  'uploads')));
app.get('/', (req, res) => {
  res.sendFile(path.join(root, '/dist/client/index.html'));
});
```

The `path` module offers many functionalities for working with the directory structures.

We use the global `__dirname` variable to get our project's root directory. The variable holds the path of the current file. Using `path.join` with `../../` and `__dirname` gives us the real root of our project.

Express.js provides the `use` function, which runs a series of commands when a given path matches. When executing this function without a path, it is executed for every request.

We use this feature to serve our static files (the avatar images) with `express.static`. They include `bundle.js` and `bundle.css`, which are created by `npm run client:build`.

In our case, first, we pass `'/'` with `express.static` following it. The result of this is that all the files and folders in `dist` are served beginning with `'/'`. The other paths in the first parameter of `app.use`, such as `'/example'`, would lead to our `bundle.js` file being able to be downloaded under `'/example/bundle.js'` instead.

For example, all the avatar images are served under `'/uploads/'`.

We are now prepared to let the client download all the necessary files. The initial route for our client is `'/'`, as specified by `app.get`. The response to this path is `index.html`. We run `res.sendFile` and the file path to return this file – that is all we have to do here.

Be sure to execute `npm run client:build` first. Otherwise, you will receive an error message, stating that these files were not found. Furthermore, when running `npm run client`, the `dist` folder is deleted, so you must rerun the build process.

Refreshing your browser now will present you with the *post* feed and form from *Chapter 1, Preparing Your Development Environment*.

The next section focuses on the great functionality of middleware functions in Express.js.

Using Express.js middleware

Express.js provides us with great ways to write efficient backends without duplicating code.

Every middleware function receives a request, a response, and `next`. It needs to run `next` to pass control to the next handler function. Otherwise, you will receive a timeout. Middleware allows us to pre- or post-process the request or response object, execute custom code, and much more. Previously, we covered a simple example of handling requests in Express.js.

Express.js can have multiple routes for the same path and HTTP method. The middleware can decide which function should be executed.

The following code is an easy example that shows what can generally be accomplished with Express.js. You can test this by replacing the current `app.get` routes:

1. The root path, `'/'`, is used to catch any requests:

    ```
    app.get('/', function (req, res, next) {
    ```

2. Here, we will randomly generate a number with `Math.random` between 1 and 10:

    ```
    var random = Math.random() * (10 -1) + 1;
    ```

3. If the number is higher than 5, we run the `next('route')` function to skip to the next `app.get` with the same path:

    ```
    if (random > 5) next('route')
    ```

    This route will log `'second'`.

4. If the number is lower than `0.5`, we execute the `next` function without any parameters and go to the next handler function. This handler will log `'first'`:

    ```
    else next()
    }, function (req, res, next) {
    res.send('first');
    })

    app.get('/', function (req, res, next) {
    res.send('second');
    })
    ```

You do not need to copy this code as it is just an explanatory example. This functionality can come in handy when we cover special treatments such as admin users and error handling.

Installing important middleware

For our application, we have already used one built-in Express.js middleware: `express.static`. Throughout this book, we will continue to install other pieces of middleware:

```
npm install --save compression cors helmet
```

Now, add the `import` statement for the new packages inside the server's `index.js` file so that all the dependencies are available within the file:

```
import helmet from 'helmet';
import cors from 'cors';
import compress from 'compression';
```

Let's see what these packages do and how we can use them.

Express Helmet

Helmet is a tool that allows you to set various HTTP headers to secure your application.

We can enable the Express.js Helmet middleware as follows in the server's `index.js` file. Add the following code snippet directly beneath the `app` variable:

```
app.use(helmet());
app.use(helmet.contentSecurityPolicy({
  directives: {
    defaultSrc: ["'self'"],
    scriptSrc: ["'self'", "'unsafe-inline'"],
    styleSrc: ["'self'", "'unsafe-inline'"],
    imgSrc: ["'self'", "data:", "*.amazonaws.com"]
  }
}));
app.use(helmet.referrerPolicy({ policy: 'same-origin' }));
```

We are doing multiple things here at once. In the preceding code, we added some **cross-site scripting** (**XSS**) protection tactics and removed the `X-Powered-By` HTTP header, as well as some other useful things, just by using the `helmet()` function in the first line.

> **Note**
>
> You can look up the default parameters, as well as the other functionalities of Helmet, at `https://github.com/helmetjs/helmet`. Always be conscious when implementing security features and do your best to verify your attack protection methods.

Furthermore, to ensure that no one can inject malicious code, we used the `Content-Security-Policy` HTTP header or CSP for short. This header prevents attackers from loading resources from external URLs.

As you can see, we also specified the `imgSrc` field, which tells our client that only images from these URLs should be loaded, including **Amazon Web Services** (**AWS**). We will learn how to upload images to it in *Chapter 7, Handling Image Uploads*.

You can read more about CSP and how it can make your platform more secure at `https://helmetjs.github.io/docs/csp/`.

The last enhancement is to set the `Referrer` HTTP header, but only when making requests on the same host. When we're going from domain A to domain B, for example, we do not include the referrer, which is the URL the user is coming from. This enhancement stops any internal routing or requests from being exposed to the internet.

It is important to initialize Helmet very high in your Express router so that all the responses are affected.

Compression with Express.js

Enabling compression for Express.js saves you and your user bandwidth, and this is easy to do. The following code must also be added to the server's `index.js` file:

```
app.use(compress());
```

This middleware compresses all the responses going through it. Remember to add it very high in your routing order so that all the requests are affected.

> **Note**
>
> Whenever you have middleware like this or multiple routes that match the same path, you need to check the initialization order. Only the first matching route is executed unless you run the `next` command. All the routes that are defined afterward will not be executed.

CORS in Express.js

We want our GraphQL API to be accessible from any website, app, or system. A good idea might be to build an app or offer the API to other companies or developers so that they can use it. When you're using APIs via Ajax, the main problem is that the API needs to send the correct `Access-Control-Allow-Origin` header.

For example, if you build the API, publicize it under `https://api.example.com`, and try to access it from `https://example.com` without setting the correct header, it won't work. The API would need to set at least `example.com` inside the `Access-Control-Allow-Origin` header to allow this domain to access its resources. This seems a bit tedious, but it makes your API open to cross-site requests, which you should always be aware of.

Allow **cross-origin resource sharing (CORS)** requests by adding the following command to the `index.js` file:

```
app.use(cors());
```

This command handles all the problems we usually have with cross-origin requests at once. It merely sets a wildcard with * inside of `Access-Control-Allow-Origin`, allowing anyone from anywhere to use your API, at least in the first instance. You can always secure your API by offering API keys or by only allowing access to logged-in users. Enabling CORS only allows the requesting site to receive the response.

Furthermore, the command also implements the `OPTIONS` route for the whole application.

The `OPTIONS` method or request is made every time we use `CORS`. This action is what's called a **preflight** request, which ensures that the responding server trusts you. If the server does not respond correctly to the `OPTIONS` preflight, the actual method, such as `POST`, will not be made by the browser at all.

Our application is now ready to serve all the routes appropriately and respond with the right headers.

Now, let's set up a GraphQL server.

Combining Express.js with Apollo

First things first; we need to install the Apollo and GraphQL dependencies:

```
npm install --save apollo-server-express graphql @graphql-
tools/schema
```

Apollo offers an Express.js-specific package that integrates itself into the web server. There is also a standalone version without Express.js. Apollo allows you to use the available Express.js middleware. In some scenarios, you may need to offer non-GraphQL routes to proprietary clients who do not implement GraphQL or are not able to understand JSON responses. There are still reasons to offer some fallbacks to GraphQL. In those cases, you can rely on Express.js since you are already using it.

Create a separate folder for services. A service can be GraphQL or other routes:

```
mkdir src/server/services/
mkdir src/server/services/graphql
```

Create an index.js file in the graphql folder to act as the start point for our GraphQL service. It must handle multiple things for initialization. Let's go through all of them one by one and add them to the index.js file:

1. First, we must import the apollo-server-express and @graphql-tools/ schema packages:

    ```
    import { ApolloServer } from 'apollo-server-express';
    import { makeExecutableSchema } from '@graphql-tools/
    schema';
    ```

2. Next, we must combine the GraphQL schema with the resolver functions. We must import the corresponding schema and resolver functions at the top from separate files. The GraphQL schema is the representation of the API – that is, the data and functions a client can request or run. Resolver functions are the implementation of the schema. Both need to match. You cannot return a field or run a mutation that is not inside the schema:

    ```
    import Resolvers from './resolvers';
    import Schema from './schema';
    ```

3. The makeExecutableSchema function of the @graphql-tools/schema package merges the GraphQL schema and the resolver functions, resolving the data we are going to write. The makeExecutableSchema function throws an error when you define a query or mutation that is not in the schema. The resulting schema is executed by our GraphQL server resolving the data or running the mutations we request:

    ```
    const executableSchema = makeExecutableSchema({
      typeDefs: Schema,
      resolvers: Resolvers
    });
    ```

4. We pass this as a schema parameter to the Apollo Server. The `context` property contains the `request` object of Express.js. In our resolver functions, we can access the request if we need to:

```
const server = new ApolloServer({
    schema: executableSchema,
    context: ({ req }) => req
});
```

5. This `index.js` file exports the initialized server object, which handles all GraphQL requests:

```
export default server;
```

Now that we are exporting the Apollo Server, it needs to be imported somewhere else. I find it convenient to have one `index.js` file on the services layer so that we only rely on this file if a new service is added.

Create an `index.js` file in the `services` folder and enter the following code:

```
import graphql from './graphql';

export default {
    graphql,
};
```

The preceding code requires our `index.js` file from the `graphql` folder and re-exports all the services into one big object. We can define more services here if we need them.

To make our GraphQL server publicly accessible to our clients, we are going to bind the Apollo Server to the `/graphql` path.

Import the services `index.js` file into the `server/index.js` file, as follows:

```
import services from './services';
```

The `services` object only holds the `graphql` index. Now, we must bind the GraphQL server to the Express.js web server with the following code:

```
const serviceNames = Object.keys(services);

for (let i = 0; i < serviceNames.length; i += 1) {
    const name = serviceNames[i];
```

```
  if (name === 'graphql') {
    (async () => {
      await services[name].start();
      services[name].applyMiddleware({ app });
    })();
  } else {
    app.use('/${name}', services[name]);
  }
}
```

For convenience, we loop through all indexes of the `services` object and use the index as the name of the route the service will be bound to. The path would be `/example` for the `example` index in the `services` object. For a typical service, such as a REST interface, we rely on the standard `app.use` method of Express.js.

Since the Apollo Server is kind of special, when binding it to Express.js, we need to run the `applyMiddleware` function, which is provided by the initialized Apollo Server, and avoid using the `app.use` function from Express.js. Apollo automatically binds itself to the `/graphql` path because it is the default option. You could also include a `path` parameter if you want it to respond from a custom route.

The Apollo Server requires us to run the `start` command before applying the middleware. As this is an asynchronous function, we are wrapping the complete block into a wrapping `async` function so that we can use the `await` statement.

Two things are missing now: the schema and the resolvers. Once we've done that, we will execute some test GraphQL requests. The schema is next on our to-do list.

Writing your first GraphQL schemas

Let's start by creating a `schema.js` inside the `graphql` folder. You can also stitch multiple smaller schemas into one bigger schema. This would be cleaner and would make sense when your application, types, and fields grow. For this book, one file is okay and we can insert the following code into the `schema.js` file:

```
const typeDefinitions = '
  type Post {
    id: Int
    text: String
  }
```

```
  type RootQuery {
    posts: [Post]
  }

  schema {
    query: RootQuery
  }
';

export default [typeDefinitions];
```

The preceding code represents a basic schema, which would be able to at least serve the fake posts array from *Chapter 1, Preparing Your Development Environment*, excluding the users.

First, we must define a new type called `Post`. A `Post` type has an `id` of `Int` and a `text` value of `String`.

For our GraphQL server, we need a type called `RootQuery`. The `RootQuery` type wraps all of the queries a client can run. It can be anything from requesting all posts, all users, posts by just one user, and so on. You can compare this to all `GET` requests as you find them with common REST APIs. The paths would be `/posts`, `/users`, and `/users/ ID/posts` to represent the GraphQL API as a REST API. When using GraphQL, we only have one route, and we send the query as a JSON-like object.

The first query we will have is going to return an array of all of the posts we have got.

If we query for all posts and want to return each user with its corresponding post, this would be a sub-query that would not be represented in our `RootQuery` type but in the `Post` type itself. You will see how this is done later.

At the end of the JSON-like schema, we add `RootQuery` to the `schema` property. This type is the starting point for the Apollo Server.

Later, we are going to add the mutation key to the schema, where we will implement a `RootMutation` type. It is going to serve all of the actions a user can run. Mutations are comparable to the `POST`, `UPDATE`, `PATCH`, and `DELETE` requests of a REST API.

At the end of the file, we export the schema as an array. If we wanted to, we could push other schemas to this array to merge them.

The last thing that's missing here is the implementation of our resolvers.

Implementing GraphQL resolvers

Now that the schema is ready, we need the matching resolver functions.

Create a `resolvers.js` file in the `graphql` folder, as follows:

```
const resolvers = {
  RootQuery: {
    posts(root, args, context) {
      return [];
    },
  },
};


export default resolvers;
```

The `resolvers` object holds all types as a property. Here, we set up `RootQuery`, holding the `posts` query in the same way as we did in our schema. The `resolvers` object must equal the schema but be recursively merged. If you want to query a subfield, such as the user of a post, you have to extend the `resolvers` object with a `Post` object containing a `user` function next to `RootQuery`.

If we send a query for all posts, the `posts` function is executed. There, you can do whatever you want, but you need to return something that matches the schema. So, if you have an array of `posts` as the response type of `RootQuery`, you cannot return something different, such as just one post object instead of an array. In that case, you would receive an error.

Furthermore, GraphQL checks the data type of every property. If `id` is defined as `Int`, you cannot return a regular MongoDB `id` since these IDs are of the `String` type. GraphQL would throw an error too.

> **Note**
>
> GraphQL will parse or cast specific data types for you if the value type is matching. For example, a `string` with a value of `2.1` is parsed to `Float` without any problems. On the other hand, an empty string cannot be converted into a `Float`, and an error would be thrown. It is better to directly have the correct data types because this saves you from casting and also prevents unwanted problems.

To prove that everything is working, we will continue by performing a real GraphQL request against our server. Our `posts` query will return an empty array, which is a correct response for GraphQL. We will come back to the `resolver` functions later. You should be able to start the server again so that we can send a demo request.

Sending GraphQL queries

We can test this query using any HTTP client, such as Postman, Insomnia, or any you are used to. The next section covers HTTP clients. If you want to send the following queries on your own, then you can read the next section and come back here.

You can test our new function when you send the following JSON as a `POST` request to `http://localhost:8000/graphql`:

```
{
  "operationName": null,
  "query": "{
    posts {
      id
      text
    }
  }",
  "variables": {}
}
```

The `operationName` field is not required to run a query, but it is great for logging purposes.

The `query` object is a JSON-like representation of the query we want to execute. In this example, we run the `RootQuery` posts and request the `id` and `text` fields of every post. We do not need to specify `RootQuery` because it is the highest layer of our GraphQL API.

The `variables` property can hold parameters such as the user IDs that we want to filter the posts by. If you want to use variables, they need to be defined in the query by their name too.

For developers who are not used to tools such as Postman, there is also the option to open the /graphql endpoint in a separate browser tab. You will be presented with a GraphQLi instance that's made for sending queries easily. Here, you can insert the content of the query property and hit the play button. Because we set up Helmet to secure our application, we need to deactivate it in development. Otherwise, the GraphQLi instance is not going to work. Just wrap the complete Helmet initialization inside the server/index.js file with the following if statement in curly braces:

```
if (process.env.NODE_ENV === 'production')
```

This short condition only activates Helmet when the environment is in development. Now, you can send the request with GraphQLi or any HTTP client.

The response of the POST request, when combined with the preceding body, should look as follows:

```
{
    "data": {
      "posts": []
    }
}
```

Here, we received the empty posts array, as expected.

Going further, we want to respond with the fake data we statically wrote in our client so that it comes from our backend. Copy the initialPosts array from App.js above the resolvers object but rename it posts. We can respond to the GraphQL request with this filled posts array.

Replace the content of the posts function in the GraphQL resolvers with this:

```
return posts;
```

You can rerun the POST request and receive both fake posts. The response does not include the user object we have in our fake data, so we must define a user property on the post type in our schema to fix this issue.

Using multiple types in GraphQL schemas

Let's create a User type and use it with our posts. First, add it to the schema:

```
type User {
  avatar: String
  username: String
}
```

Now that we have a User type, we need to use it inside the Post type. Add it to the Post type, as follows:

```
user: User
```

The user field allows us to have a sub-object inside our posts, along with the post's author information.

Our extended query to test this looks like this:

```
"query":"{
  posts {
    id
    text
    user {
      avatar
      username
    }
  }
}"
```

You cannot just specify the user as a property of the query. Instead, you need to provide a sub-selection of fields. This is required whenever you have multiple GraphQL types nested inside each other. Then, you need to select the fields your result should contain.

Running the updated query gives us the fake data, which we already have in our frontend code; just the posts array as-is.

We have made good progress with querying data, but we also want to be able to add and change data.

Writing your first GraphQL mutation

One thing our client has already offered was adding new posts to the fake data temporarily. We can implement this in the backend by using GraphQL mutations.

Starting with the schema, we need to add the mutation, as well as the input types, as follows:

```
input PostInput {
  text: String!
}

input UserInput {
  username: String!
  avatar: String!
}

type RootMutation {
  addPost (
    post: PostInput!
    user: UserInput!
  ): Post
}
```

GraphQL inputs are no more than types. Mutations can use them as parameters inside requests. They may look weird because our current output types look almost the same. However, it would be wrong to have an id property on PostInput, for example, since the backend chooses the ID and the client cannot provide it. Consequently, it does make sense to have separate objects for input and output types.

The addPost function receiving our two new required input types – PostInput and UserInput – is a new feature. Those functions are called mutations since they mutate the current state of the application. The response to this mutation is an ordinary Post object. When creating a new post with the addPost mutation, we will directly get the created post from the backend in response.

The exclamation mark in the schema tells GraphQL that the field is a required parameter.

The RootMutation type corresponds to the RootQuery type and is an object that holds all GraphQL mutations.

The last step is to enable the mutations in our schema for the Apollo Server by applying the `RootMutation` type to the `schema` object:

```
schema {
  query: RootQuery
  mutation: RootMutation
}
```

> **Note**
>
> Usually, the client does not send the user with the mutation. This is because the user is authenticated first, before adding a post, and through that, we already know which user initiated the Apollo request. However, we will ignore this for the moment and implement authentication later in *Chapter 6, Authentication with Apollo and React*.

Now, the `addPost` resolver function needs to be implemented in the `resolvers.js` file.

Add the following `RootMutation` object to `RootQuery` in `resolvers.js`:

```
RootMutation: {
  addPost(root, { post, user }, context) {
    const postObject = {
      ...post,
      user,
      id: posts.length + 1,
    };
    posts.push(postObject);
    return postObject;
  },
},
```

This resolver extracts the `post` and `user` objects from the mutation's parameters, which are passed in the second argument of the function. Then, we build the `postObject` variable. We want to add our `posts` array as a property by destructuring the `post` input and adding the `user` object. The `id` field is just the length of the `posts` array plus one.

Now, the `postObject` variable looks like a `post` from the `posts` array. Our implementation does the same as the frontend is already doing. The return value of our `addPost` function is `postObject`. To get this working, you need to change the initialization of the `posts` array from `const` to `let`. Otherwise, the array will be static and unchangeable.

You can run this mutation via your preferred HTTP client, like so:

```
{
    "operationName": null,
    "query": "mutation addPost($post : PostInput!,
      $user: UserInput!) {
      addPost(post : $post, user: $user) {
        id
        text
        user {
          username
          avatar
        }
      }
    }",
    "variables": {
      "post": {
        "text": "You just added a post."
      },
      "user": {
        "avatar": "/uploads/avatar3.png",
        "username": "Fake User"
      }
    }
}
```

First, we pass the word `mutation` and the actual function name – in this case, `addPost` – that we want to run, including a selection of response fields inside the `query` property, to the normal data query for the posts.

Second, we use the `variables` property to send the data we want to insert into our backend. We need to pass them as parameters within the `query` string. We can define both parameters with a dollar sign and the awaited data type inside the `operation` string. The variables marked with dollar signs are then mapped into the actual action we want to trigger on the backend.

When we send this mutation, the request will have a `data` object, including an `addPost` field. The `addPost` field holds the post, which we send with our request.

If you query the posts again, you will see that there are now three posts. Great – it worked!

As with our client, this is only temporary until we restart the server. We'll cover how to persist data in a SQL database in *Chapter 3, Connecting to the Database*.

Next, we'll cover the various ways you can debug your backend.

Backend debugging and logging

Two things are very important for debugging. The first is that we need to implement logging for our backend in case we receive errors from our users, while the second is that we need to look into Postman to debug our GraphQL API efficiently.

So, let's get started with logging.

Logging in Node.js

The most popular logging package for Node.js is called `winston`. Install and configure `winston` by following these steps:

1. Install `winston` with npm:

    ```
    npm install --save winston
    ```

2. Next, create a new folder for all of the helper functions from the backend:

    ```
    mkdir src/server/helpers
    ```

3. Then, insert a `logger.js` file into the new folder with the following content:

    ```
    import winston from 'winston';

    let transports = [
      new winston.transports.File({
        filename: 'error.log',
        level: 'error',
      }),
      new winston.transports.File({
        filename: 'combined.log',
        level: 'verbose',
      }),
    ];
    ```

```
if (process.env.NODE_ENV !== 'production') {
  transports.push(new winston.transports.Console());
}

const logger = winston.createLogger({
  level: 'info',
  format: winston.format.json(),
  transports,
});

export default logger;
```

This file can be imported everywhere we want to log.

In the preceding code, we defined the standard `transports` for `winston`. A transport is nothing more than how `winston` separates and saves various log types in different files.

The first `transport` generates an `error.log` file where only real errors are saved.

The second transport is a combined log where we save all the other log messages, such as warnings or information logs.

If we are running the server in a development environment, which we are currently doing, we must add a third transport. We will also directly log all the messages to the console while developing on the server.

Most people who are used to JavaScript development know the difficulty with `console.log`. By directly using `winston`, we can see all the messages in the Terminal, but we do not need to clean the code from `console.log` either, so long as the things we log make sense.

To test this out, we can try the `winston` logger in the only mutation we have.

In `resolvers.js`, add the following code to the top of the file:

```
import logger from '../../helpers/logger';
```

Now, we can extend the `addPost` function by adding the following before the `return` statement:

```
logger.log({ level: 'info', message: 'Post was created' });
```

When you send the mutation now, you will see that the message was logged to the console.

Furthermore, if you look in the root folder of your project, you will see the `error.log` and `combined.log` files. The `combined.log` file should contain the log from the console.

Now that we can log all the operations on the server, we should explore Postman so that we can send requests comfortably.

Debugging with Postman

Postman is one of the most widely used HTTP clients there is. It not only provides raw HTTP client functionality but also teams and collections, along with letting you synchronize all the requests you saved in Postman.

You can install Postman by downloading the appropriate file from `https://www.postman.com/downloads/`.

> **Note**
> Numerous other HTTP client tools are useful for debugging your application. You are free to use your tool of choice. Some other great clients that I use are Insomnia, SoapUI, and Stoplight, but there are many more. In this book, we will use Postman, as it is the most popular from my point of view.

When you have finished the installation, it should look something like this:

Figure 2.1 – Postman screen after installing the Book collection

As you can see, I have already created a collection called **Book** in the left-hand panel. This collection includes our two requests: one to request all posts and one to add a new post.

As an example, the following screenshot shows you what the **Add Post** mutation looks like in Postman:

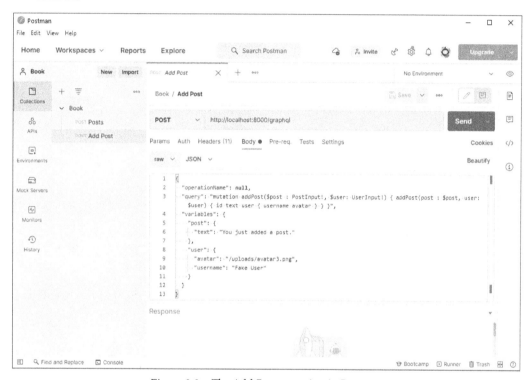

Figure 2.2 – The Add Post mutation in Postman

The URL is localhost and includes port 8000, as expected.

The request body looks pretty much like what we saw previously. Be sure to select application/json as Content-Type next to the raw format.

> **Note**
>
> In my case, I need to write the query inline because Postman is not able to handle multi-row text inside JSON. If this is not the case for you, please ignore it.

Since the newer version of Postman was released, there is also the option to select GraphQL instead of JSON. If you do that, you can write the GraphQL code in multiple lines and write the variables in a separate window. The result should look like this:

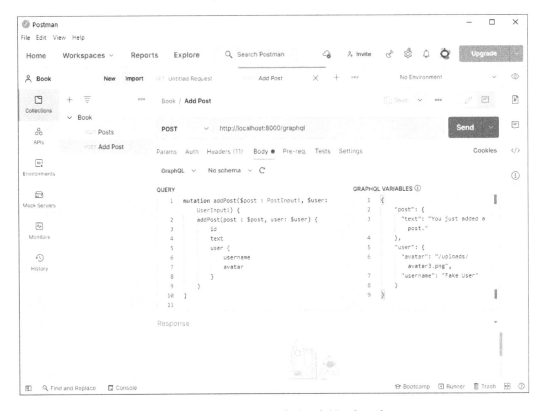

Figure 2.3 – Postman with GraphQL selected

If you add a new request, you can use the *Ctrl + S* shortcut to save it. You need to select a collection and a name to save it with. One major downfall of using Postman (at least with GraphQL APIs) is that we are only using POST. It would be great to have some kind of indication of what we are doing here – for example, a query or a mutation. We will learn how to use authorization in Postman once we have implemented it.

Postman also has other great features, such as automated testing, monitoring, and mocking a fake server.

Later in this book, it will become more complicated to configure Postman for all requests. In such cases, I like to use the Apollo Client Developer Tools, which perfectly integrate into the frontend and make use of Chrome DevTools. What's great about the Apollo Client Developer Tools is that they use the Apollo Client we configure in the frontend code, which means they reuse the authentication we built into our frontend.

Summary

In this chapter, we set up our Node.js server with Express.js and bound the Apollo Server to respond to requests on a GraphQL endpoint. We can handle queries, return fake data, and mutate that data with GraphQL mutations.

Furthermore, we can log every process in our Node.js server. Debugging an application with Postman leads to a well-tested API, which can be used later in our frontend.

In the next chapter, we will learn how to persist data in a SQL server. We will also implement models for our GraphQL types and cover migrations for our database. We need to replace our current `resolver` functions with queries via Sequelize.

There is a lot to do, so read on for more!

3

Connecting to the Database

Our backend and frontend can communicate, create new posts, and respond with a list of all posts while using fake data. The next step on our list will be to use a database, such as a SQL server, to serve as data storage.

We want our backend to persist data to our SQL database by using Sequelize. Our Apollo Server should use this data for queries and mutations, as needed. For this to happen, we must implement database models for our GraphQL entities.

This chapter will cover the following topics:

- Using databases in GraphQL
- Using Sequelize in Node.js
- Writing database models
- Seeding data with Sequelize
- Using Apollo with Sequelize
- Performing database migrations with Sequelize

Technical requirements

The source code for this chapter is available in the following GitHub repository:
`https://github.com/PacktPublishing/Full-Stack-Web-Development-with-GraphQL-and-React-Second-Edition/tree/main/Chapter03`.

Using databases in GraphQL

GraphQL is a protocol for sending and receiving data. **Apollo** is one of the many libraries that you can use to implement that protocol. Neither GraphQL (in its specifications) nor Apollo works directly on the data layer. Where the data that you put into your response comes from, and where the data that you send with your request is saved, are up to the developer to decide.

This logic indicates that the database and the services that you use do not matter to Apollo, so long as the data that you respond with matches the GraphQL schema.

As we are living in the Node.js ecosystem in this project and book, it would be fitting to use MongoDB. MongoDB offers a great client library for Node.js and uses JavaScript as its native choice of language for interactions and querying.

The general alternative to a database system such as MongoDB is a typical MySQL server with proven stability and global usage. One case that I encounter frequently involves systems and applications relying on older code bases and databases that need upgrades. A great way to accomplish this is to get an over-layering API level with GraphQL. In this scenario, the GraphQL server receives all requests and, one by one, you can replace the existing code bases that the GraphQL server relies on. In these cases, it is helpful that GraphQL is database agnostic.

In this book, we will use SQL via Sequelize to see this feature in a real-world use case. For future purposes, it will also help you to handle problems with existing SQL-based systems.

Installing MySQL for development

MySQL is an excellent starting point for getting on track in a developmental career. It is also well suited for local development on your machine since the setup is easy.

How to set up MySQL on your machine depends on your operating system. As we mentioned in *Chapter 1*, *Preparing Your Development Environment*, we are assuming that you are using a Debian-based system. For this, you can use the following instructions. If you already have a working setup for MySQL or Apache, these commands may not work, or may not be required in the first place.

> **Tip**
>
> For other operating systems, there are great prebuilt packages. I recommend that all Windows users use XAMPP and that Mac users use MAMP. These offer an easy installation process for what we did manually on Linux. They also implement MySQL, Apache, and PHP, including phpMyAdmin.
>
> **Important Note**
>
> Do not follow these instructions when setting up a real SQL server for public and production use. A professional setup includes many security features to protect you against attacks. This installation should only be used in development, on your local machine.

Execute the following steps to get MySQL running:

1. First, you should always install all the updates available for your system:

   ```
   sudo apt-get update && sudo apt-get upgrade -y
   ```

   We want to install MySQL and a GUI to see what we have inside our database. The most common GUI for a MySQL server is phpMyAdmin. For this, you need to install a web server and PHP. We are going to install Apache as our web server.

   > **Important Note**
   >
   > If, at any point in the process, you receive an error stating that the package could not be found, ensure that your system is Debian-based. The installation process can differ on other systems. You can easily search for the matching package for your system on the internet.

2. Install all the necessary dependencies with the following command:

   ```
   sudo apt-get install apache2 mysql-server php php-pear
   php-mysql
   ```

3. After the installation, you will need to run the MySQL setup in the root shell. You will have to enter the root password for this. Alternatively, you can run `sudo -i`:

   ```
   su -
   ```

4.   Now, you can execute the MySQL installation command; follow the steps as prompted:

```
mysql_secure_installation
```

You can ignore most of these steps and security settings but be careful when you are asked for the root password of your MySQL instance.

5.   We must create a separate user for development aside from the root user. You are discouraged from using the root user at all. Log into our MySQL server with the root user to accomplish this:

```
mysql -u root
```

6.   Now, run the following SQL command.

```
GRANT ALL PRIVILEGES ON *.* TO 'devuser'@'%' IDENTIFIED
BY 'PASSWORD';
```

You can replace the PASSWORD string with the password that you want. This is the password that you will use for the database connection in your application, but also when logging into phpMyAdmin. This command creates a user called devuser, with root privileges that are acceptable for local development.

> **Note**
>
> If you are already using MySQL8, the command that you need execute is a little different. Just run the following lines:
>
> ```
> CREATE USER 'devuser'@'%' IDENTIFIED BY 'PASSWORD';
> ```
>
> ```
> GRANT ALL PRIVILEGES ON *.* TO 'devuser'@'%' WITH
> GRANT OPTION;
> ```
>
> ```
> FLUSH PRIVILEGES;
> ```
>
> The above commands will create a new user with the same permissions on your MySQL server.

7.   You can install phpMyAdmin since our MySQL server has been set up. You will be asked for a web server when executing the following command. Select apache2 with the spacebar and navigate to **ok** by hitting the *Tab* key. When you're asked for it, select the automatic setup method for phpMyAdmin. You should not do this manually.

Furthermore, phpMyAdmin will want you to enter a password. I recommend that you choose the same password that you chose for the root user:

```
sudo apt-get install phpmyadmin
```

8. After the installation, we will need to set up Apache to serve phpMyAdmin. The following `ln` command creates a symbolic link in the root folder of the Apache public `HTML` folder. Apache will now serve phpMyAdmin:

```
cd /var/www/html/
sudo ln -s /usr/share/phpmyadmin
```

Now, we can visit phpMyAdmin under `http://localhost/phpmyadmin` and log in with the newly created user. This should look as follows:

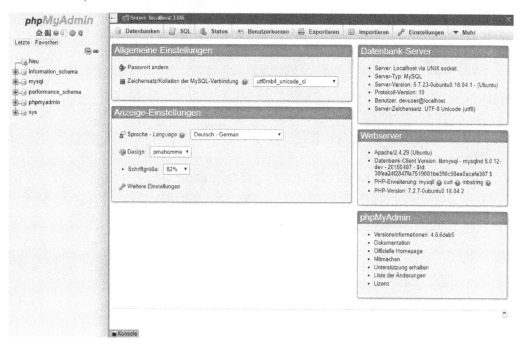

Figure 3.1 – phpMyAdmin

With that, we have installed the database for our development environment.

phpMyAdmin chooses the language according to your environment, so it might differ slightly from what's shown in the preceding screenshot.

Creating a database in MySQL

Before we start implementing our backend, we need to add a new database that we can use.

You are free to do this via the command line or phpMyAdmin. As we have just installed phpMyAdmin, we are going to use it.

You can run raw SQL commands in the **SQL** tab of phpMyAdmin. The corresponding command to create a new database is as follows:

```
CREATE DATABASE graphbook_dev CHARACTER SET utf8 COLLATE utf8_
general_ci;
```

Otherwise, you can follow the next set of steps to use the graphical method. In the left-hand panel, click on the **New** button.

You will be presented with a screen like the following. It will show all the databases, including their collation of your MySQL server:

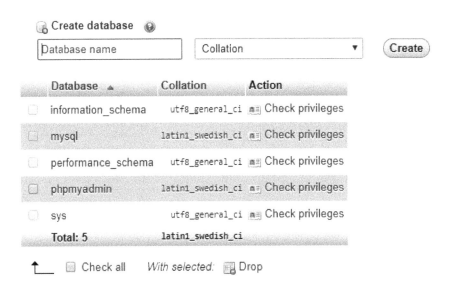

Figure 3.2 – phpMyAdmin databases

Enter a database name, such as graphbook_dev, and then choose the uft8_general_ci collation. After doing so, click on **Create**.

You will see a page that says **No tables found in database**, which is correct. This will change later, when we have implemented our database models, such as posts and users.

In the next chapter, we will start to set up Sequelize in Node.js and connect it to our SQL server.

Integrating Sequelize into our Node.js stack

We have just set up a MySQL database, and we want to use it inside of our Node.js backend. There are many libraries to connect to and query your MySQL database. We are going to use Sequelize in this book.

> **Alternative Object–Relational Mappers (ORMs)**
>
> Alternatives include Waterline ORM and js-data, which offer the same functionalities as Sequelize. What is great about these is that they not only offer SQL dialects, but also feature database adapters for MongoDB, Redis, and more. So, if you need an alternative, check them out.

Sequelize is an ORM for Node.js. It supports the PostgreSQL, MySQL, SQLite, and MSSQL standards.

Install Sequelize in your project via npm. We will also install a second package, called mysql2:

```
npm install --save sequelize mysql2
```

The mysql2 package allows Sequelize to speak with our MySQL server.

Sequelize is just a wrapper around the various libraries for the different database systems. It offers great features for intuitive model usage, as well as functions for creating and updating database structures and inserting development data.

Typically, you would run npx sequelize-cli init before starting with the database connection or models, but I prefer a more custom approach. From my point of view, this is a bit cleaner. This approach is also why we are setting up the database connection in an extra file and are not relying on boilerplate code.

> **Setting Up Sequelize Traditionally**
>
> You can look at the official tutorial in the Sequelize documentation if you want to see how it would usually be done. The approach that we are taking and the one from the tutorial do not differ too much, but it is always good to see another way of doing things. The documentation can be found at https://sequelize.org/master/manual/migrations.html.

Let's start by setting Sequelize up in our backend.

Connecting to a database with Sequelize

The first step is to initialize the connection from Sequelize to our MySQL server. To do this, we will create a new folder and file, as follows:

```
mkdir src/server/database
touch src/server/database/index.js
```

Inside the `index.js` database, we will establish a connection to our database with Sequelize. Internally, Sequelize relies on the `mysql2` package, but we do not use it ourselves, which is very convenient:

```
import Sequelize from 'sequelize';

const sequelize = new Sequelize('graphbook_dev', 'devuser',
'PASSWORD', {
  host: 'localhost',
  dialect: 'mysql',
  pool: {
    max: 5,
    min: 0,
    acquire: 30000,
    idle: 10000,
  },
});

export default sequelize;
```

As you can see, we load Sequelize from `node_modules` and then create an instance of it. The following properties are important for Sequelize:

- We pass the database name as the first parameter, which we just created.
- The second and third parameters are the credentials of our `devuser`. Replace them with the username and password that you entered for your database. `devuser` has permission to access all the databases in our MySQL server. This makes development a lot easier.
- The fourth parameter is a general options object that can hold many more properties. The preceding object is an example configuration.

- The `host` option of our MySQL database is our local machine alias, `localhost`. If this is not the case, you can also specify the IP or URL of the MySQL server.

- `dialect` is, of course, `mysql`.

- With the `pool` option, you tell Sequelize the configuration for every database connection. The preceding configuration allows for a minimum of zero connections, which means that Sequelize should not maintain one connection, but should create a new one whenever it is needed. The maximum number of connections is five. This option also relates to the number of replica sets that your database system has.

  The `idle` field of the `pool` option specifies how long a connection can be unused before it gets closed and removed from the pool of active connections.

  When trying to establish a new connection to our MySQL server, the timeout before the connection is aborted is defined by the `acquire` option. In cases in which a connection cannot be created, this option helps stop your server from freezing.

Executing the preceding code will instantiate Sequelize and will successfully create a connection to our MySQL server. Going further, we need to handle multiple databases for every environment that our application can run in, from development to production. You will see that in the next section.

Using a configuration file with Sequelize

The previous setup for our database connection with Sequelize is fine, but it is not made for later deployment. The best option is to have a separate configuration file that is read and used according to the environment that the server is running in.

For this, create a new `index.js` file inside a separate folder (called `config`), next to the `database` folder:

```
mkdir src/server/config
touch src/server/config/index.js
```

Your sample configuration should look like the following code if you have followed the instructions for creating a MySQL database. The only thing that we did here was copy our current configuration into a new object indexed with the `development` or `production` environment:

```
module.exports = {
  "development": {
    "username": "devuser",
    "password": "PASSWORD",
```

```
    "database": "graphbook_dev",
    "host": "localhost",
    "dialect": "mysql",
    "pool": {
      "max": 5,
      "min": 0,
      "acquire": 30000,
      "idle": 10000
    }
  },
  "production": {
    "host": process.env.host,
    "username": process.env.username,
    "password": process.env.password,
    "database": process.env.database,
    "logging": false,
    "dialect": "mysql",
    "pool": {
      "max": 5,
      "min": 0,
      "acquire": 30000,
      "idle": 10000
    }
  }
}
```

Sequelize expects a `config.json` file inside this folder by default, but this setup will allow us to take on a more custom approach in later chapters. The `development` environment directly stores the credentials for your database, whereas the `production` configuration uses environment variables to fill them.

We can remove the configuration that we hardcoded earlier and replace the contents of our `database/index.js` file to require our `configFile` instead.

This should look as follows:

```
import Sequelize from 'sequelize';
import configFile from '../config/';
```

```
const env = process.env.NODE_ENV || 'development';
const config = configFile[env];

const sequelize = new Sequelize(config.database,
  config.username, config.password, config);

const db = {
  sequelize,
};

export default db;
```

In the preceding code, we are using the NODE_ENV environment variable to get the environment that the server is running in. We read the config file and pass the correct configuration to the Sequelize instance. The environment variable will allow us to switch to a new environment, such as production, later in this book.

The Sequelize instance is then exported for use throughout our application. We use a special db object for this. You will see why we are doing this later.

Next, you will learn how to generate and write models and migrations for all the entities that our application will have.

Writing database models

After creating a connection to our MySQL server via Sequelize, we want to use it. However, our database is missing a table or structure that we can query or manipulate. Creating those is the next thing that we need to do.

Currently, we have two GraphQL entities: User and Post.

Sequelize lets us create a database schema for each of our GraphQL entities. The schema is validated when we insert or update rows in our database. We already wrote a schema for GraphQL in the schema.js file, which is used by Apollo Server, but we need to create a second one for our database. The field types, as well as the fields themselves, can vary between the database and the GraphQL schema.

GraphQL schemas can have more fields than our database model, or vice versa. Perhaps you do not want to export all the data from your database through the API, or maybe you want to generate data for your GraphQL API on the fly when you're requesting data.

Let's create the first model for our posts. Create two new folders (one called `models` and another called `migrations`) next to the `database` folder:

```
mkdir src/server/models
mkdir src/server/migrations
```

Creating each model in a separate file is much cleaner than having one big file for all the models.

Your first database model

We will use the Sequelize CLI to generate our first database model. Install it globally with the following command:

```
npm install -g sequelize-cli
```

This gives you the ability to run the `sequelize` command inside your Terminal.

The Sequelize CLI allows us to generate the model automatically. This can be done by running the following command:

```
sequelize model:generate --models-path src/server/models
--migrations-path src/server/migrations --name Post
--attributes text:text
```

Sequelize expects us to run the command in the folder where we have run `sequelize init` by default. Our file structure is different because we have two layers with `src/server`. For this reason, we specify the path manually with the first two parameters; that is, `--models-path` and `--migrations-path`.

The `--name` parameter gives our model a name under which it can be used. The `--attributes` option specifies the fields that the model should include.

> **Tip**
> If you are customizing your setup, you may want to know about other options that the CLI offers. You can view the manual for every command easily by appending `--help` as an option: `sequelize model:generate --help`.

This command creates a `post.js` model file in your `models` folder and a database migration file, named `XXXXXXXXXXXXXX-create-post.js`, in your `migrations` folder. The X icons indicate the timestamp when you're generating the files with the CLI. You will see how migrations work in the next section.

The following model file was created for us:

```
'use strict';
const {
  Model
} = require('sequelize');
module.exports = (sequelize, DataTypes) => {
  class Post extends Model {
    /**
     * Helper method for defining associations.
     * This method is not a part of Sequelize lifecycle.
     * The 'models/index' file will call this method
       automatically.
     */
    static associate(models) {
      // define association here
    }
  };
  Post.init({
    text: DataTypes.TEXT
  }, {
    sequelize,
    modelName: 'Post',
  });
  return Post;
};
```

Here, we are creating the Post class and extending the Model class from Sequelize. Then, we are using the init function of Sequelize to create a database model:

- The first parameter is the model attributes.

- The second parameter is an option object, where the sequelize connection instance and model name are passed.

> **Model Customization**
>
> There are many more options that Sequelize offers us to customize our database models. If you want to look up which options are available, you can find them at `https://sequelize.org/master/manual/model-basics.html`.

A `post` object has the `id`, `text`, and `user` properties. The user will be a separate model, as shown in the GraphQL schema. Consequently, we only need to configure `id` and `text` as columns of a post.

`id` is the key that uniquely identifies a data record from our database. We do not specify this when running the `model:generate` command because it is generated by MySQL automatically.

The `text` column is just a MySQL `TEXT` field that allows us to write long posts. Alternatively, there are other MySQL field types, with `MEDIUMTEXT`, `LONGTEXT`, and `BLOB`, which could save more characters. A regular `TEXT` column should be fine for our use case.

The Sequelize CLI created a model file, exporting a function that, after execution, returns the real database model. You will soon see why this is a great way of initializing our models.

Let's take a look at the migration file that is also created by the CLI.

Your first database migration

So far, MySQL has not known anything about our plan to save posts inside of it. Our database tables and columns need to be created, hence why the migration file was created.

A migration file has multiple advantages, such as the following:

1. Migrations allow us to track database changes through our regular version control system, such as Git or SVN. Every change to our database structure should be covered in a migration file.
2. Migration files enable us to write updates that automatically apply database changes for new versions of our application.

Our first migration file creates a `Posts` table and adds all the required columns, as follows:

```
'use strict';

module.exports = {
```

```
up: async (queryInterface, Sequelize) => {
  await queryInterface.createTable('Posts', {
    id: {
      allowNull: false,
      autoIncrement: true,
      primaryKey: true,
      type: Sequelize.INTEGER
    },
    text: {
      type: Sequelize.TEXT
    },
    createdAt: {
      allowNull: false,
      type: Sequelize.DATE
    },
    updatedAt: {
      allowNull: false,
      type: Sequelize.DATE
    }
  });
},
down: async (queryInterface, Sequelize) => {
  await queryInterface.dropTable('Posts');
}
};
```

By convention, the model name is pluralized in migrations, but it is singular inside model definitions. Our table names are also pluralized. Sequelize offers options to change this.

A migration has two properties, as follows:

- The up property states what should be done when running the migration.
- The down property states what is run when undoing a migration.

As we mentioned previously, the id and text columns are created, as well as two additional datetime columns, to save the creation and update time.

The id field has set autoIncrement and primaryKey to true. id will count upward, from one to nearly infinite, for each post in our table. This id uniquely identifies posts for us. Passing allowNull with false disables this feature so that we can insert a row with an empty field value.

To execute this migration, we will use the Sequelize CLI again, as follows:

```
sequelize db:migrate --migrations-path src/server/migrations
--config src/server/config/index.js
```

Look inside of phpMyAdmin. Here, you will find the new table, called posts. The structure of the table should look as follows:

#	Name	Type	Collation	Attributes	Null	Default	Comments	Extra
1	id	int(11)			No	None		AUTO_INCREMENT
2	text	text	utf8_general_ci		Yes	NULL		
3	createdAt	datetime			No	None		
4	updatedAt	datetime			No	None		

Figure 3.3 – Posts table structure

All the columns were created as we desired.

Furthermore, two additional fields – createdAt and updatedAt – were created. These two fields tell us when a row was either created or updated. The fields were created by Sequelize automatically. If you do not want this, you can set the timestamps property in the model to false.

Every time you use Sequelize and its migration feature, you will have an additional table called sequelizemeta. The contents of the table should look as follows:

Figure 3.4 – Migrations table

Sequelize saves every migration that has been executed. If we add more fields in development or a news release cycle, we can write a migration that runs all the table-altering statements for us as an update. Sequelize skips all the migrations that are saved inside the meta table.

One major step is to bind our model to Sequelize. This process can be automated by running `sequelize init`, but understanding it will teach us way more than relying on premade boilerplate commands.

Importing models with Sequelize

We want to import all the database models at once, in a central file. Our database connection instantiator will then rely on this file.

Create an `index.js` file in the `models` folder and use the following code:

```
import Sequelize from 'sequelize';
if (process.env.NODE_ENV === 'development') {
  require('babel-plugin-require-context-hook/register')()
}

export default (sequelize) => {
  let db = {};

  const context = require.context('.', true,
    /^\.\/(?!index\.js).*\.js$/, 'sync')
  context.keys().map(context).forEach(module => {
    const model = module(sequelize, Sequelize);
    db[model.name] = model;
  });

  Object.keys(db).forEach((modelName) => {
    if (db[modelName].associate) {
      db[modelName].associate(db);
    }
  });

  return db;
};
```

The preceding logic will also be generated when running `sequelize init`, but this way, the database connection is set up in a separate file from loading the models. Usually, this would happen in just one file when using the Sequelize boilerplate code. Furthermore, we have introduced some webpack-specific configurations.

To summarize what happens in the preceding code, we search for all the files ending with
.js in the same folder as the current file and load them all with the require.context
statement. In development, we must execute the babel-plugin-require-context-
hook/register Hook to load the require.context function at the top. This
package must be installed with npm, with the following command:

```
npm install --save-dev babel-plugin-require-context-hook
```

We need to load the Babel plugin at the start of our development server, so, open the
package.json file and edit the server script, as follows:

```
nodemon --exec babel-node --plugins require-context-hook
--watch src/server src/server/index.js
```

When the plugin is loaded and we run the require('babel-plugin-require-
context-hook/register')() function, the require.context method becomes
available for us. Make sure that you set the NODE_ENV variable to development;
otherwise, this won't work.

In production, the require.context function is included in the generated bundle
of webpack.

The loaded model files export a function with the following two parameters:

- Our Sequelize instance, after creating a connection to our database
- The sequelize class itself, including the data types it offers, such as integer or text

Running the exported functions imports the actual Sequelize model. Once all the models
have been imported, we loop through them and check whether they have a function
called associate. If this is the case, we execute the associate function and, through
that, we establish relationships between multiple models. Currently, we have not set up an
association, but that will change later in this chapter.

Now, we want to use our models. Go back to the index.js database file and import all
the models through the aggregated index.js file that we just created:

```
import models from '../models';
```

Before exporting the db object at the end of the file, we need to run the models wrapper
to read all model .js files. We pass our Sequelize instance as a parameter, as follows:

```
const db = {
  models: models(sequelize),
```

```
    sequelize,
};
```

The new database object in the preceding command has `sequelize` and `models` as properties. Under `models`, you can find the `Post` model and every new model that we are going to add later.

The database `index.js` file is ready and can now be used. You should import this file only once because it can get messy when you're creating multiple instances of Sequelize. The pool functionality won't work correctly, and we will end up with more connections than the maximum of five that we specified earlier.

We must create the global database instance in the `index.js` file of the root server folder. Add the following code:

```
import db from './database';
```

We require the `database` folder and the `index.js` file inside this folder. Loading the file instantiates the Sequelize object, including all the database models.

Going forward, we want to query some data from our database via the GraphQL API that we implemented in *Chapter 2, Setting Up GraphQL with Express.js*.

Seeding data with Sequelize

We should fill the empty `Posts` table with our fake data. To accomplish this, we will use Sequelize's feature for seeding data to our database.

Create a new folder called `seeders`:

```
mkdir src/server/seeders
```

Now, we can run our next Sequelize CLI command to generate a boilerplate file:

```
sequelize seed:generate --name fake-posts --seeders-path src/
server/seeders
```

Seeders are great for importing test data into a database for development. Our `seed` file has the timestamp and the words `fake-posts` in the name, and should look as follows:

```
'use strict';

module.exports = {
  up: (queryInterface, Sequelize) => {
```

```
  /*
    Add altering commands here.
    Return a promise to correctly handle asynchronicity.

    Example:
    return queryInterface.bulkInsert('Person', [{
      name: 'John Doe',
      isBetaMember: false
    }], {});
  */
  },
  down: (queryInterface, Sequelize) => {
  /*
    Add reverting commands here.
    Return a promise to correctly handle asynchronicity.

    Example:
    return queryInterface.bulkDelete('Person', null, {});
  */
  }
};
```

As shown in the preceding code snippet, nothing is done here. It is just an empty boilerplate file. We need to edit this file to create the fake posts that we already had in our backend. This file looks like our migration from the previous section. Replace the contents of the file with the following code:

```
'use strict';

module.exports = {
  up: (queryInterface, Sequelize) => {
    return queryInterface.bulkInsert('Posts', [{
      text: 'Lorem ipsum 1',
      createdAt: new Date(),
      updatedAt: new Date(),
    },
    {
```

```
        text: 'Lorem ipsum 2',
        createdAt: new Date(),
        updatedAt: new Date(),
      }],
      {});
    },
    down: (queryInterface, Sequelize) => {
      return queryInterface.bulkDelete('Posts', null, {});
    }
};
```

In the up migration, we are bulk inserting two posts through queryInterface and its bulkInsert command. For this, we will pass an array of posts, excluding the id property and the associated user. This id is created automatically, and the user is saved in a separate table later. queryInterface of Sequelize is the general interface that Sequelize uses to talk to all databases.

In our seed file, we need to add the createdAt and updatedAt fields since Sequelize does not set up default values for the timestamp columns in MySQL. In reality, Sequelize takes care of the default values of those fields by itself, but not when seeding data. If you do not provide these values, the seed will fail, because NULL is not allowed for createdAt and updatedAt.

The down migration bulk deletes all the rows in the table since this is the apparent reverse action of the up migration.

Execute all the seeds from the seeders folder with the following command:

```
sequelize db:seed:all --seeders-path src/server/seeders
--config src/server/config/index.js
```

Sequelize does not check or save whether a seed has been run already, as we are doing it with the preceding command. This means that you can run seeds multiple times if you want to.

The following screenshot shows a filled-in Posts table:

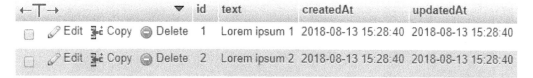

	id	text	createdAt	updatedAt
✏ Edit ⏳ Copy ⊖ Delete	1	Lorem ipsum 1	2018-08-13 15:28:40	2018-08-13 15:28:40
✏ Edit ⏳ Copy ⊖ Delete	2	Lorem ipsum 2	2018-08-13 15:28:40	2018-08-13 15:28:40

Figure 3.5 – The Posts table with seed data

The demo posts are now inside our database.

We will cover how to use Sequelize with our Apollo Server, as well as how to add the relationship between the user and their posts, in the next section.

Using Sequelize with GraphQL

The database object is initialized upon starting the server within the root `index.js` file. We pass it from this global location down to the spots where we rely on the database. This way, we do not import the database file repeatedly but have a single instance that handles all the database queries for us.

The services that we want to publicize through the GraphQL API need access to our MySQL database. The first step is to implement the posts in our GraphQL API. It should respond with the fake posts from the database we just inserted.

Global database instance

To pass the database down to our GraphQL resolvers, we must create a new object in the server `index.js` file:

```
import db from './database';

const utils = {
    db,
};
```

Here, we created a `utils` object directly under the `import` statement of the `database` folder.

The `utils` object holds all the utilities that our services might need access to. This can be anything from third-party tools to our MySQL server, or any other database, as shown in the preceding code.

Replace the line where we import the `services` folder, as follows:

```
import servicesLoader from './services';
const services = servicesLoader(utils);
```

The preceding code might look weird to you, but what we are doing here is executing the function that is the result of the `import` statement and passing the `utils` object as a parameter. We must do this in two separate lines since the `import` syntax does not allow it in just one line; so, we must first import the function that's been exported from the `services` folder into a separate variable.

So far, the return value of the `import` statement has been a simple object. We have to change this so that it matches our requirements.

To do this, go to the services `index.js` file and change the contents of the file, as follows:

```
import graphql from './graphql';

export default utils => ({
  graphql: graphql(utils),
});
```

We surrounded the preceding `services` object with a function, which was then exported. That function accepts only one parameter, which is our `utils` object.

That object is then given to a new function, called `graphql`. Every service that we are going to use has to be a function that accepts this parameter. This allows us to hand over any property that we want to the deepest point in our application.

When executing the preceding exported function, the result is the regular `services` object we used previously. We only wrapped it inside a function to pass the `utils` object.

The `graphql` import that we are doing needs to accept the `utils` object.

Open the `index.js` file from the `graphql` folder and replace everything but the `require` statements at the top with the following code:

```
export default (utils) => {
    const server = new ApolloServer({
        typeDefs: Schema,
        resolvers: Resolvers.call(utils),
        context: ({ req }) => req
    });

    return server;
};
```

Again, we have surrounded everything with a function that accepts the `utils` object. The aim of all this is to have access to the database within our GraphQL resolvers, which are given to `ApolloServer`.

To accomplish this, we are using the `Resolvers.call` function from JavaScript. This function allows us to set the owner object of the exported `Resolvers` function. What we are saying here is that the scope of `Resolvers` is the `utils` object.

So, within the `Resolvers` function, accessing this now gives us the `utils` object as the scope. At the moment, `Resolvers` is just a simple object, but because we used the `call` method, we must also return a function from the `resolvers.js` file.

Surround the `resolvers` object in this file with a function and return the `resolvers` object from inside the function:

```
export default function resolver() {
    ...
    return resolvers;
}
```

We cannot use the arrow syntax, as we did previously. ES6 arrow syntax would automatically take a scope, but we want the `call` function to take over here.

An alternative way of doing this would be to hand over the `utils` object as a parameter. I think the way that we have chosen to do things is a bit cleaner, but handle it as you like.

Running the first database query

Now, we can start to use the database. Add the following code to the top of the `export default function resolver` statement:

```
const { db } = this;
const { Post } = db.models;
```

The `this` keyword is the owner of the current method and holds the `db` object, as stated previously. We extracted the database models from the `db` object that we built in the previous section.

The good thing about models is that you do not need to write raw queries against the database. You have already told Sequelize which fields and tables it can use by creating a model. At this point, you can use Sequelize's methods to run queries against the database within your resolvers.

We can query all the posts through the Sequelize model, instead of returning the fake posts from before. Replace the `posts` property within `RootQuery` with the following code:

```
posts(root, args, context) {
  return Post.findAll({order: [['createdAt', 'DESC']]});
},
```

In the preceding code, we searched for and selected all the posts that we have in our database. We used the Sequelize `findAll` method and returned the result of it. The return value will be a JavaScript promise, which automatically gets resolved once the database has finished collecting the data.

A typical news feed, such as on Twitter or Facebook, orders the posts according to the creation date. That way, you have the newest posts at the top and the oldest at the bottom. Sequelize expects an array of arrays as a parameter of the order property that we pass as the first parameter to the `findAll` method. The results are ordered by their creation date.

> **Important Note**
>
> There are many other methods that Sequelize offers. You can query for just one entity, count them, find them, create them if they are not found, and much more. You can look up the methods that Sequelize provides at `https://sequelize.org/master/manual/model-querying-basics.html`.

You can start the server with `npm run server` and execute the GraphQL posts query from *Chapter 2, Setting Up GraphQL with Express.js*, again. The output will look as follows:

```
{
  "data": {
    "posts": [{
      "id": 1,
      "text": "Lorem ipsum 1",
      "user": null
    },
    {
      "id": 2,
      "text": "Lorem ipsum 2",
      "user": null
    }]
  }
}
```

The `id` and `text` fields look fine, but the `user` object is `null`. This happened because we did not define a user model or declare a relationship between the user and the post model. We will change this in the next section.

One-to-one relationships in Sequelize

We need to associate each post with a user, to fill in the gap that we created in our GraphQL response. A post must have an author. It would not make sense to have a post without an associated user.

First, we will generate a `User` model and migration. We will use the Sequelize CLI again, as follows:

```
sequelize model:generate --models-path src/server/models
--migrations-path src/server/migrations --name User
--attributes avatar:string,username:string
```

The migration file creates the `Users` table and adds the `avatar` and `username` columns. A data row looks like a post in our fake data, but it also includes an autogenerated ID and two timestamps, as you saw previously.

The relationship of the users to their specific posts is still missing as we have only created the model and migration file. We still have to add the relationship between posts and users. This will be covered in the next section.

What every post needs is an extra field called `userId`. This column acts as the foreign key to reference a unique user. Then, we can join the user that's related to each post.

> **Note**
> MySQL offers great documentation for people that are not used to foreign key constraints. If you are one of them, you should read up on this topic at `https://dev.mysql.com/doc/refman/8.0/en/create-table-foreign-keys.html`.

Updating the table structure with migrations

We have to write a third migration, adding the `userId` column to our `Post` table, but also including it in our database `Post` model.

Generating a boilerplate migration file is very easy with the Sequelize CLI:

```
sequelize migration:create --migrations-path src/server/
migrations --name add-userId-to-post
```

You can directly replace the content of the generated migration file, as follows:

```
'use strict';

module.exports = {
  up: async (queryInterface, Sequelize) => {
    return Promise.all([
      queryInterface.addColumn('Posts',
        'userId',
        {
          type: Sequelize.INTEGER,
        }),
      queryInterface.addConstraint('Posts', {
        fields: ['userId'],
        type: 'foreign key',
        name: 'fk_user_id',
        references: {
          table: 'Users',
          field: 'id',
        },
        onDelete: 'cascade',
        onUpdate: 'cascade',
      }),
    ]);
  },

  down: async (queryInterface, Sequelize) => {
    return Promise.all([
      queryInterface.removeColumn('Posts', 'userId'),
    ]);
  }
};
```

This migration is a bit more complex, and I will explain it on a step-by-step basis:

1. In the up migration, we are using `queryInterface` to add the `userId` column to the `Posts` table.

2. Next, we are adding a foreign key constraint with the `addConstraint` function. This constraint represents the relationship between both the user and the post entities. The relationship is saved in the `userId` column of the `Post` table.

3. I experienced some issues when running the migrations without using `Promise.all`, which ensures that all the promises in the array are resolved. Returning only the array did not run both the `addColumn` and `addConstraint` methods.

4. The preceding `addConstraint` function receives the `foreign key` string as a `type`, which says that the data type is the same as the corresponding column in the `Users` table. We want to give our constraint the custom name `fk_user_id` to identify it later.

5. Then, we are specifying the `references` field for the `userId` column. Sequelize requires a table, which is the `Users` table, and the field that our foreign key relates to, which is the `id` column of the `User` table. This is everything that is required to get a working database relationship.

6. Furthermore, we are changing the `onUpdate` and `onDelete` constraints to `cascade`. What this means is that, when a user either gets deleted or has their user ID updated, the change is reflected in the user's posts. Deleting a user results in deleting all the posts of a user, while updating a user's ID updates the ID on all the user's posts. We do not need to handle all this in our application code, which would be inefficient.

> **Note**
> There is a lot more about this topic in the Sequelize documentation. If you want to read up on this, you can find more information at `https://sequelize.org/master/manual/query-interface.html`.

Rerun the migration to see what changes occurred:

```
sequelize db:migrate --migrations-path src/server/migrations
--config src/server/config/index.js
```

The benefit of running migrations through Sequelize is that it goes through all the possible migrations from the `migrations` folder. It excludes those that are already saved inside the `SequelizeMeta` table, and then chronologically runs the migrations that are left. Sequelize can do this because the timestamp is included in every migration's filename.

After running the migration, there should be a `Users` table, and the `userId` column should have been added to the `Posts` table.

Take a look at the relationship view of the `Posts` table in phpMyAdmin. You can find it under the **Structure** view, by clicking on **Relation view**:

Figure 3.6 – MySQL foreign keys

As you can see, we have our foreign key constraint. The correct name was taken, as well as the cascade option.

If you receive an error when running migrations, you can easily undo them, as follows:

```
sequelize db:migrate:undo --migrations-path src/server/
migrations --config src/server/config/index.js
```

This command undoes the most recent migrations. Always be conscious of what you do here. Keep a backup if you are unsure whether everything works correctly.

You can also revert all migrations at once, or only revert to one specific migration so that you can go back to a specific timestamp:

```
sequelize db:migrate:undo:all --to XXXXXXXXXXXXXX-create-posts.
js --migrations-path src/server/migrations --config src/server/
config/index.js
```

Leave out the `--to` parameter to undo all migrations.

With that, we have established the database relationship, but Sequelize must know about the relationship too. You will learn how this is done in the next section.

Model associations in Sequelize

Now that we have the relationship configured with the foreign key, it needs to be configured inside our Sequelize model.

Go back to the `Post` model file and replace the `associate` function with the following code:

```
static associate(models) {
  this.belongsTo(models.User);
}
```

The `associate` function gets evaluated inside our aggregating `index.js` file, where all the model files are imported.

We are using the `belongsTo` function here, which tells Sequelize that every post belongs to exactly one user. Sequelize gives us a new function on the `Post` model, called `getUser`, to retrieve the associated user. This naming is done by convention, as you can see. Sequelize does all this automatically.

Do not forget to add `userId` as a queryable field to the `Post` model itself, as follows:

```
userId: DataTypes.INTEGER,
```

The `User` model needs to implement the reverse association too. Add the following code to the `User` model file:

```
static associate(models) {
   this.hasMany(models.Post);
}
```

The `hasMany` function means the exact opposite of the `belongsTo` function. Every user can have multiple posts associated in the `Post` table. It can be anything, from zero to multiple posts.

You can compare the new data layout with the preceding one. Up to this point, we had the posts and users inside one big array of objects. Now, we have split every object into two tables. Both tables connect through the foreign key. This is required every time we run the GraphQL query to get all the posts, including their authors.

So, we must extend our current `resolvers.js` file. Add the `Post` property to the `resolvers` object, as follows:

```
Post: {
  user(post, args, context) {
    return post.getUser();
  },
},
```

`RootQuery` and `RootMutation` were the two main properties that we've had so far. `RootQuery` is the starting point where all GraphQL queries begin.

With the old demo posts, we were able to directly return a valid and complete response, since everything that we needed was in there already. Now, a second query, or a `JOIN`, is being executed to collect all the necessary data for a complete response.

The `Post` entity is introduced to our `resolvers`, where we can define functions for every property of our GraphQL schema. Only the user is missing in our response; the rest is there. That is why we have added the `user` function to the resolvers.

The first parameter of the function is the `post` model instance that we are returning inside the `RootQuery` resolver.

Then, we are using the `getUser` function that Sequelize gave us. Executing the `getUser` function runs the correct MySQL `SELECT` query to get the correct user from the `Users` table. It does not run a real MySQL `JOIN`; it only queries the user in a separate MySQL command. Later, in the *Chats and messages in GraphQL* section, you will learn about another way to run a `JOIN` directly, which is more efficient.

However, if you query for all the posts via the GraphQL API, the user will still be `null`. We have not added any users to the database yet, so let's insert them next.

Seeding foreign key data

The challenge of adding users is that we have already introduced a foreign key constraint to the database. You can follow these instructions to learn how to get it working:

1. First, we must use the Sequelize CLI to generate an empty `seeders` file, as follows:

    ```
    sequelize seed:generate --name fake-users --seeders-path
    src/server/seeders
    ```

2. Fill in the following code to insert the fake users:

    ```
    'use strict';

    module.exports = {
      up: async (queryInterface, Sequelize) => {
        return queryInterface.bulkInsert('Users', [{
          avatar: '/uploads/avatar1.png',
          username: 'TestUser',
          createdAt: new Date(),
          updatedAt: new Date(),
        },
        {
          avatar: '/uploads/avatar2.png',
          username: 'TestUser2',
          createdAt: new Date(),
    ```

```
      updatedAt: new Date(),
    }],
    {});
  },
  down: async (queryInterface, Sequelize) => {
    return queryInterface.bulkDelete('Users', null,
      {});
  }
};
```

The preceding code looks like the `seeders` file for the posts, but instead, we are now inserting users with the correct fields. Every user receives an auto-incremented ID by our MySQL server when inserting a user.

3. We must maintain the relationships that have been configured in our database. Adjust the `posts` seed file to reflect this and replace the up migration so that the correct user IDs are inserted for every post:

```
up: (queryInterface, Sequelize) => {
  // Get all existing users
  return queryInterface.sequelize.query(
    'SELECT id from Users;',
  ).then((users) => {
    const usersRows = users[0];
    return queryInterface.bulkInsert('Posts', [{
      text: 'Lorem ipsum 1',
      userId: usersRows[0].id,
      createdAt: new Date(),
      updatedAt: new Date(),
    },
    {
      text: 'Lorem ipsum 2',
      userId: usersRows[1].id,
      createdAt: new Date(),
      updatedAt: new Date(),
    }],
    {});
  });
},
```

Here, we are using a raw MySQL query to get all the users and their IDs to insert them with our posts. This ensures that we have a valid foreign key relationship that MySQL allows us to insert.

The posts we have currently stored in our table do not receive a `userId`, and we do not want to write a separate migration or seed to fix those posts.

There are two options here. You can either manually truncate the tables through phpMyAdmin and SQL statements, or you can use the Sequelize CLI. It is easier to use the CLI, but the result will be the same either way. The following command will undo all the seeds:

```
sequelize db:seed:undo:all --seeders-path src/server/seeders
--config src/server/config/index.js
```

When undoing seeds, the tables are not truncated, so the `autoIncrement` index is not set back to one; instead, it stays at the current index. Reverting seeds multiple times raises the user's or post's ID, which stops the seeds from working. We have fixed this by using the raw MySQL query that retrieves the current user IDs before inserting the posts.

We have one problem before running our seeders again: we created the `users` seed file after the `post` seeders file. This means that the posts are inserted before the users exist, because of the timestamps of the files. Generally, this is not a problem, but since we have introduced a foreign key constraint, we are not able to insert posts with a `userId` when the underlying user does not exist in our database. MySQL forbids this. Simply adjust the timestamp of the fake user seed file so that it's before the post seed file's timestamp, or vice versa.

After renaming the files, run all the seeds with the following command again:

```
sequelize db:seed:all --seeders-path src/server/seeders
--config src/server/config/index.js
```

If you take a look inside your database, you should see a filled `Posts` table, including `userId`. The `Users` table should look as follows:

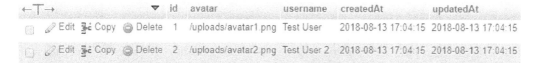

	id	avatar	username	createdAt	updatedAt
Edit Copy Delete	1	/uploads/avatar1.png	Test User	2018-08-13 17:04:15	2018-08-13 17:04:15
Edit Copy Delete	2	/uploads/avatar2.png	Test User 2	2018-08-13 17:04:15	2018-08-13 17:04:15

Figure 3.7 – The Users table

Now, you can rerun the GraphQL query, and you should see a working association between the users and their posts because the `user` field is filled in.

So far, we have achieved a lot as we can serve data from our database through the GraphQL API by matching its schema.

> **Note**
>
> There are some ways to automate this process, through additional npm packages. There is a package that automates the process of creating a GraphQL schema from your database models for you. As always, you are more flexible when you do not rely on pre-configured packages. You can find the package at `https://www.npmjs.com/package/graphql-tools-sequelize`.

Mutating data with Sequelize

Requesting data from our database via the GraphQL API works. Now comes the tough part: adding a new post to the `Posts` table.

Before we start, we must extract the new database model from the `db` object at the top of the exported function in our `resolvers.js` file:

```
const { Post, User } = db.models;
```

Currently, we have no authentication to identify the user that is creating the post. We will fake this step until the authentication is implemented *Chapter 6, Authentication with Apollo and React*.

We have to edit the GraphQL resolvers to add the new post. Replace the old `addPost` function with the new one, as shown in the following code snippet:

```
addPost(root, { post }, context) {
  return User.findAll().then((users) => {
    const usersRow = users[0];

    return Post.create({
      ...post,
    }).then((newPost) => {
      return Promise.all([
        newPost.setUser(usersRow.id),
      ]).then(() => {
```

```
        logger.log({
          level: 'info',
          message: 'Post was created',
        });
        return newPost;
      });
    });
  });
},
```

As always, the preceding mutation returns a promise. This promise is resolved when the deepest query has been executed successfully. The execution order is as follows:

1. We retrieve all the users from the database through the `User.findAll` method.

2. We insert the post into our database with the `create` function from Sequelize. The only property that we pass is the `post` object from the original request, which only holds the text of the post. MySQL autogenerates the `id` property of the post.

> **Note**
>
> Sequelize also offers a `build` function, which initializes the model instance for us. In this case, we would have to run the `save` method to insert the model manually. The `create` function does this for us all at once.

3. The post has been created, but `userId` has not been set.

    You could also directly add the user ID to the `Post.create` function. The problem here is that we would not establish the model associations, even though this is reflected in the database. If we return the created post model without explicitly using `setUser` on the model instance, we cannot use the `getUser` function, which is used to return the user for the mutation's response.

    So, to fix this problem, we must run the `create` function, resolve the promise, and then run `setUser` separately. As a parameter of `setUser`, we statically take the ID of the first user from the `users` array.

    We resolve the promise of the `setUser` function by using an array surrounded by `Promise.all`. This allows us to add further Sequelize methods later. For example, you could add a category to each post, too.

4. Once we have set `userId` correctly, the returned value is the newly created post model instance.

Everything is set now. To test our API, we are going to use Postman again. We need to change the `addPost` request. `userInput`, which we added previously, is not needed anymore, because the backend statically chooses the first user out of our database. You can send the following request body:

```
{
  "operationName": null,
  "query": "mutation addPost($post : PostInput!) {
    addPost(post : $post) {
    id text user { username avatar }}}",
  "variables":{
    "post": {
      "text": "You just added a post."
    }
  }
}
```

Your GraphQL schema must reflect this change, so remove `userInput` from there, too:

```
addPost (
  post: PostInput!
): Post
```

Running the `addPost` GraphQL mutation now adds a post to the `Posts` table, as shown in the following screenshot:

	id	text	createdAt	updatedAt	userId
Edit Copy Delete	1	Lorem ipsum 1	2018-08-14 11:08:28	2018-08-14 11:08:28	1
Edit Copy Delete	2	Lorem ipsum 2	2018-08-14 11:08:28	2018-08-14 11:08:28	2
Edit Copy Delete	3	You just added a post	2018-08-14 11:08:46	2018-08-14 11:08:46	1

Figure 3.8 – Post inserted into the database table

As we are not using the demo `posts` array anymore, you can remove it from the `resolvers.js` file.

With that, we have rebuilt the example from the previous chapter, but we are using a database in our backend. To extend our application, we are going to add two new entities called `Chat` and `Message`.

Many-to-many relationships

Facebook provides users with various ways to interact. Currently, we only have the opportunity to request and insert posts. As is the case with Facebook, we want to have chats with our friends and colleagues. We will introduce two new entities to cover this.

The first entity is called `Chat`, while the second entity is called `Message`.

Before we start the implementation, we need to lay out a detailed plan of what those entities will enable us to do.

A user can have multiple chats, and a chat can belong to multiple users. This relationship allows us to have group chats with multiple users, as well as private chats between only two users. A message belongs to one user, but every message also belongs to one chat.

Model and migrations

When transferring this into real code, we must generate the `Chat` model. The problem here is that we have a many-to-many relationship between users and chats. In MySQL, this kind of relationship requires a table to store the relationships between all the entities separately.

These tables are called **join tables**. Instead of using a foreign key on the chat or a user to save the relationship, we have a table called `user_chats`. The user's ID and the chat's ID are associated with each other inside this table. If a user participates in multiple chats, they will have multiple rows in this table, with different chat IDs.

Chat model

Let's start by creating the `Chat` model and migration. A chat itself does not store any data; we use it for grouping specific users' messages:

```
sequelize model:generate --models-path src/server/models
--migrations-path src/server/migrations --name Chat
--attributes firstName:string,lastName:string,email:string
```

Generate the migration for our association table, as follows:

```
sequelize migration:create --migrations-path src/server/
migrations --name create-user-chats
```

Adjust the `users_chats` migration that's generated by the Sequelize CLI. We specify the user and chat IDs as attributes for our relationship. References inside a migration automatically create foreign key constraints for us. The migration file should look as follows:

```
'use strict';

module.exports = {
  up: async (queryInterface, Sequelize) => {
    return queryInterface.createTable('users_chats', {
      id: {
        allowNull: false,
        autoIncrement: true,
        primaryKey: true,
        type: Sequelize.INTEGER
      },
      userId: {
        type: Sequelize.INTEGER,
        references: {
          model: 'Users',
          key: 'id'
        },
        onDelete: 'cascade',
        onUpdate: 'cascade',
      },
      chatId: {
        type: Sequelize.INTEGER,
        references: {
          model: 'Chats',
          key: 'id'
        },
        onDelete: 'cascade',
        onUpdate: 'cascade',
      },
      createdAt: {
        allowNull: false,
```

```
          type: Sequelize.DATE
        },
      updatedAt: {
        allowNull: false,
        type: Sequelize.DATE
      }
    });
  },
  down: async (queryInterface, Sequelize) => {
    return queryInterface.dropTable('users_chats');
  }
};
```

A separate model file for the association table is not needed because we can rely on this table in the models where the association is required. The id column could be left out because the row can only be identified by the user and chat ID.

Associate the User model with the Chat model via the new relationship table in the User model, as follows:

```
this.belongsToMany(models.Chat, { through: 'users_chats' });
```

Do the same for the Chat model, as follows:

```
this.belongsToMany(models.User, { through: 'users_chats' });
```

The through property tells Sequelize that the two models are related via the users_chats table. Normally, when you are not using Sequelize and are trying to select all users and chats merged in raw SQL, you need to maintain this association manually and join the three tables on your own. Sequelize's querying and association capabilities are so complex, so this is all done for you.

Rerun the migrations to let the changes take effect:

```
sequelize db:migrate --migrations-path src/server/migrations
--config src/server/config/index.js
```

The following screenshot shows what your database should look like now:

Table ▲	Action						Rows ❔	Type	Collation	Size	Overhead
☐ Chats	☆	🗐 Browse	Structure	Search	Insert	Empty	Drop	0	InnoDB	utf8_general_ci	16 KiB	-
☐ Posts	☆	🗐 Browse	Structure	Search	Insert	Empty	Drop	3	InnoDB	utf8_general_ci	32 KiB	-
☐ SequelizeMeta	☆	🗐 Browse	Structure	Search	Insert	Empty	Drop	5	InnoDB	utf8_unicode_ci	32 KiB	-
☐ Users	☆	🗐 Browse	Structure	Search	Insert	Empty	Drop	2	InnoDB	utf8_general_ci	16 KiB	-
☐ users_chats	☆	🗐 Browse	Structure	Search	Insert	Empty	Drop	0	InnoDB	utf8_general_ci	48 KiB	-
5 tables	Sum							10	InnoDB	utf8_general_ci	144 KiB	0 B

Figure 3.9 – Database structure

You should see two foreign key constraints in the relationship view of the `users_chats` table. The naming is done automatically:

Figure 3.10 – Foreign keys for the users_chats table

This setup was the tough part. Next up is the message entity, which is a simple one-to-one relationship. One message belongs to one user and one chat.

Message model

A message is much like a post, except that it can only be read inside a chat and is not public to everyone.

Generate the model and migration file with the CLI, as follows:

```
sequelize model:generate --models-path src/server/models
--migrations-path src/server/migrations --name Message
--attributes text:string,userId:integer,chatId:integer
```

Add the missing references to the created migration file by replacing the following properties:

```
userId: {
  type: Sequelize.INTEGER,
  references: {
    model: 'Users',
```

```
    key: 'id'
  },
  onDelete: 'SET NULL',
  onUpdate: 'cascade',
},
chatId: {
  type: Sequelize.INTEGER,
  references: {
    model: 'Chats',
    key: 'id'
  },
  onDelete: 'cascade',
  onUpdate: 'cascade',
},
```

Now, we can run the migrations again to create the `Messages` table using the `sequelize db:migrate` Terminal command.

The references also apply to our model file, where we need to use Sequelize's `belongsTo` function to get all those convenient model methods for our resolvers. Replace the `associate` function of the `Message` model with the following code:

```
static associate(models) {
  this.belongsTo(models.User);
  this.belongsTo(models.Chat);
}
```

In the preceding code, we defined that every message is related to exactly one user and chat.

On the other hand, we must also associate the `Chat` model with the messages. Add the following code to the `associate` function of the `Chat` model:

```
this.hasMany(models.Message);
```

The next step is to adjust our GraphQL API to provide chats and messages.

Chats and messages in GraphQL

So far, we have introduced some new entities with messages and chats. Let's include those in our Apollo schema. In the following code, you can see an excerpt of the changed entities, fields, and parameters of our GraphQL schema:

```
type User {
  id: Int
  avatar: String
  username: String
}

type Post {
  id: Int
  text: String
  user: User
}

type Message {
  id: Int
  text: String
  chat: Chat
  user: User
}

type Chat {
  id: Int
  messages: [Message]
  users: [User]
}

type RootQuery {
  posts: [Post]
  chats: [Chat]
}
```

Take a look at the following short changelog of our GraphQL schema:

- The User type received an id field, thanks to our database.

- The Message type is entirely new. It has a text field like a typical message, and user and chat fields, which are requested from the referenced tables in the database model.

- The Chat type is also new. A chat contains a list of messages that are returned as an array. These can be queried through the chat ID, which is saved in the message table. Furthermore, a chat has an unspecified number of users. The relationships between users and chats are saved in our separate **join table**. The interesting thing here is that our schema does not know anything about this table; it is just for our internal use to save the data appropriately in our MySQL server.

- I have also added a new RootQuery, called chats. This query returns all the user's chats.

These factors should be implemented in our resolvers too. Our resolvers should look as follows:

```
Message: {
  user(message, args, context) {
    return message.getUser();
  },
  chat(message, args, context) {
    return message.getChat();
  },
},
Chat: {
  messages(chat, args, context) {
    return chat.getMessages({ order: [['id', 'ASC']] });
  },
  users(chat, args, context) {
    return chat.getUsers();
  },
},
RootQuery: {
  posts(root, args, context) {
    return Post.findAll({order: [['createdAt', 'DESC']]});
  },
```

```
    chats(root, args, context) {
      return User.findAll().then((users) => {
        if (!users.length) {
          return [];
        }

        const usersRow = users[0];

        return Chat.findAll({
          include: [{
            model: User,
            required: true,
            through: { where: { userId: usersRow.id } },
          },
          {
            model: Message,
          }],
        });
      });
    },
  },
```

Let's go through the changes one by one:

- We added the Message property to our resolvers.

- We added the Chat property to the resolvers object. There, we run the
  getMessages and getUsers functions, to retrieve all the joined data. All the
  messages are sorted by the ID in ascending order (to show the latest message at
  the bottom of a chat window, for example).

- I added the new RootQuery, called chats, to return all the fields, as in
  our schema:

  a) Until we get a working authentication, we will statically use the first user when
  querying for all chats.

  b) We are using the findAll method of Sequelize and joining the users of any
  returned chat. For this, we use the include property of Sequelize on the User
  model within the findAll method. It runs a MySQL JOIN, not a second
  SELECT query.

c) Setting the `include` statement to `required` runs an `INNER JOIN`, not a `LEFT OUTER JOIN`, by default. Any chat that does not match the condition in the `through` property is excluded. In our example, the condition is that the user ID must match.

d) Lastly, we join all the available messages for each chat in the same way, without any condition.

We must use the new models here. We should not forget to extract them from the `db.models` object inside the `resolver` function. It must look as follows:

```
const { Post, User, Chat, Message } = db.models;
```

You can send this GraphQL request to test the changes:

```
{
  "operationName":null,
  "query": "{ chats { id users { id } messages { id text
    user { id username } } } }",
  "variables":{}
}
```

The response should give us an empty `chats` array, as follows:

```
{
  "data": {
    "chats": []
  }
}
```

This empty array was returned because we do not have any chats or messages in our database. You will learn how to fill it with data in the next section.

Seeding many-to-many data

Testing our implementation requires data to be in our database. We have three new tables, so we will create three new seeders to get some test data to work with.

Let's start with the chats, as follows:

```
sequelize seed:generate --name fake-chats --seeders-path src/
server/seeders
```

Now, replace the new seeder file with the following code. Running the following code creates a chat in our database. We do not need more than two timestamps because the chat ID is generated automatically:

```
'use strict';

module.exports = {
  up: async (queryInterface, Sequelize) => {
    return queryInterface.bulkInsert('Chats', [{
      createdAt: new Date(),
      updatedAt: new Date(),
    }],
    {});
  },
  down: async (queryInterface, Sequelize) => {
    return queryInterface.bulkDelete('Chats', null, {});
  }
};
```

Next, we must insert the relationship between two users and the new chat. We can do this by creating two entries in the `users_chats` table where we reference them. Now, generate the boilerplate seed file, as follows:

```
sequelize seed:generate --name fake-chats-users-relations
--seeders-path src/server/seeders
```

Our seed should look much like the previous ones, as follows:

```
'use strict';

module.exports = {
  up: async (queryInterface, Sequelize) => {
    const usersAndChats = Promise.all([
      queryInterface.sequelize.query(
        'SELECT id from Users;',
      ),
      queryInterface.sequelize.query(
        'SELECT id from Chats;',
      ),
```

```
    ]);

    return usersAndChats.then((rows) => {
      const users = rows[0][0];
      const chats = rows[1][0];

      return queryInterface.bulkInsert('users_chats', [{
        userId: users[0].id,
        chatId: chats[0].id,
        createdAt: new Date(),
        updatedAt: new Date(),
      },
      {
        userId: users[1].id,
        chatId: chats[0].id,
        createdAt: new Date(),
        updatedAt: new Date(),
      }],
      {});
    });
  },
  down: async (queryInterface, Sequelize) => {
    return queryInterface.bulkDelete('users_chats', null, {});
  }
};
```

Inside the up migration, we resolve all the users and chats using `Promise.all`. This ensures that, when the promise is resolved, all the chats and users are available at the same time. To test the chat functionality, we choose the first chat and the first two users that are returned from the database. We take their IDs and save them in our `users_chats` table. Those two users should be able to talk to each other through this one chat later.

The last table without any data in it is the `Messages` table. Generate the seed file, as follows:

```
sequelize seed:generate --name fake-messages --seeders-path
src/server/seeders
```

Again, replace the generated boilerplate code, as follows:

```
'use strict';

module.exports = {
  up: async (queryInterface, Sequelize) => {
    const usersAndChats = Promise.all([
      queryInterface.sequelize.query(
        'SELECT id from Users;',
      ),
      queryInterface.sequelize.query(
        'SELECT id from Chats;',
      ),
    ]);

    return usersAndChats.then((rows) => {
      const users = rows[0][0];
      const chats = rows[1][0];

      return queryInterface.bulkInsert('Messages', [{
        userId: users[0].id,
        chatId: chats[0].id,
        text: 'This is a test message.',
        createdAt: new Date(),
        updatedAt: new Date(),
      },
      {
        userId: users[1].id,
        chatId: chats[0].id,
        text: 'This is a second test message.',
        createdAt: new Date(),
        updatedAt: new Date(),
      }],
      {});
    });
  },
```

```
  down: async (queryInterface, Sequelize) => {
    return queryInterface.bulkDelete('Messages', null, {});
  }
};
```

Now, all the seed files should be ready. It makes sense to empty all the tables before running the seeds so that you can work with clean data. I like to delete all the tables in the database from time to time and rerun all the migrations and seeds to test them from zero. Whether or not you are doing this, you should at least be able to run the new seed.

Try to run the GraphQL chats query again. It should look as follows:

```
{
  "data": {
    "chats": [{
      "id": 1,
      "users": [
        {
          "id": 1
        },
        {
          "id": 2
        }
      ],
      "messages": [
        {
          "id": 1,
          "text": "This is a test message.",
          "user": {
            "id": 1,
            "username": "Test User"
          }
        },
        {
          "id": 2,
          "text": "This is a second test message.",
          "user": {
            "id": 2,
```

```
            "username": "Test User 2"
          }
        }
      ]}
    ]
  }
}
```

Great! Now, we can request all the chats that a user participates in and get all the referenced users and their messages.

Now, we also want to do this for only one chat. Follow these steps:

1. Add a `RootQuery` chat that takes a `chatId` as a parameter:

    ```
    chat(root, { chatId }, context) {
      return Chat.findByPk(chatId, {
        include: [{
          model: User,
          required: true,
        },
        {
          model: Message,
        }],
      });
    },
    ```

    With this implementation, we have the problem that all the users can send a query to our Apollo Server and, in return, get the complete chat history, even if they are not referenced in the chat. We will not be able to fix this until we implement authentication later in *Chapter 6, Authentication with Apollo and React*.

2. Add the new query to the GraphQL schema, under `RootQuery`:

    ```
    chat(chatId: Int): Chat
    ```

3. Send the GraphQL request to test the implementation, as follows:

    ```
    {
      "operationName":null,
      "query": "query($chatId: Int!){ chat(chatId:
      $chatId) {
    ```

```
      id users { id } messages { id text user { id
         username } } } }",
   "variables":{ "chatId": 1 }
}
```

Here, we are sending this query, including `chatId` as a parameter. To pass a parameter, you must define it in the query with its GraphQL data type. Then, you can set it in the specific GraphQL query that you are executing, which is the `chat` query. Lastly, you must insert the parameter's value into the `variables` field of the GraphQL request.

You may remember the response from the last time. The new response will look much like a result of the `chats` query, but instead of an array of chats, we will just have one `chat` object.

We are missing a major feature: sending new messages or creating a new chat. We will create the corresponding schema, and the resolvers for it, in the next section.

Creating a new chat

New users want to chat with their friends, so creating a new chat is essential.

The best way to do this is to accept a list of user IDs so that we can allow group chats too. Do this as follows:

1. Add the `addChat` function to `RootMutation` in the `resolvers.js` file, as follows:

```
addChat(root, { chat }, context) {
  return Chat.create().then((newChat) => {
    return Promise.all([
      newChat.setUsers(chat.users),
    ]).then(() => {
      logger.log({
        level: 'info',
        message: 'Message was created',
      });
      return newChat;
    });
  });
},
```

Sequelize added the `setUsers` function to the chat model instance. It was added because of the associations using the `belongsToMany` method in the chat model. There, we can directly provide an array of user IDs that should be associated with the new chat, through the `users_chats` table.

2. Change the schema so that you can run the GraphQL mutation. We must add the new input type and mutation, as follows:

```
input ChatInput {
    users: [Int]
}


type RootMutation {
    addPost (
        post: PostInput!
    ): Post
    addChat (
        chat: ChatInput!
    ): Chat
}
```

3. Test the new GraphQL `addChat` mutation as your request body:

```
{
    "operationName":null,
    "query": "mutation addChat($chat: ChatInput!) {
        addChat(chat: $chat) { id users { id } }}",
    "variables":{
        "chat": {
            "users": [1, 2]
        }
    }
}
```

You can verify that everything worked by checking the users that were returned inside the `chat` object.

Creating a new message

We can use the `addPost` mutation as our basis and extend it. The result accepts a `chatId` and uses the first user from our database. Later, the authentication will be the source of the user ID:

1. Add the `addMessage` function to `RootMutation` in the `resolvers.js` file, as follows:

```
addMessage(root, { message }, context) {
  return User.findAll().then((users) => {
    const usersRow = users[0];

    return Message.create({
      ...message,
    }).then((newMessage) => {
      return Promise.all([
        newMessage.setUser(usersRow.id),
        newMessage.setChat(message.chatId),
      ]).then(() => {
        logger.log({
          level: 'info',
          message: 'Message was created',
        });
        return newMessage;
      });
    });
  });
},
```

2. Then, add the new mutation to your GraphQL schema. We also have a new input type for our messages:

```
input MessageInput {
  text: String!
  chatId: Int!
}

type RootMutation {
  addPost (
```

```
        post: PostInput!
    ): Post
    addChat (
        chat: ChatInput!
    ): Chat
    addMessage (
        message: MessageInput!
    ): Message
}
```

3. You can send the request in the same way as the `addPost` request:

```
{
    "operationName":null,
    "query": "mutation addMessage($message :
      MessageInput!) {
      addMessage(message : $message) { id text }}",
    "variables":{
      "message": {
        "text": "You just added a message.",
        "chatId": 1
      }
    }
}
```

Now, everything is set. The client can now request all posts, chats, and messages. Furthermore, users can create new posts, create new chat rooms, and send chat messages.

Summary

Our goal in this chapter was to create a working backend with a database as storage, which we have achieved pretty well. We can add further entities and migrate and seed them with Sequelize. Migrating our database changes won't be a problem for us when it comes to going into production.

In this chapter, we also covered what Sequelize automates for us when using its models, and how great it works in coordination with our Apollo Server.

In the next chapter, we will focus on how to use the Apollo React Client library with our backend, as well as the database behind it.

Section 2: Building the Application

Knowing how to write a React application and tying that together with GraphQL and other technologies to build a real use case is the hardest part. This section will give you the confidence to build an application yourself while teaching you how React, authentication, routing, server-side rendering, and many other things work.

In this section, there are the following chapters:

- *Chapter 4, Hooking Apollo into React*
- *Chapter 5, Reusable React Components and React Hooks*
- *Chapter 6, Authentication with Apollo and React*
- *Chapter 7, Handling Image Uploads*
- *Chapter 8, Routing in React*
- *Chapter 9, Implementing Server-Side Rendering*
- *Chapter 10, Real-Time Subscriptions*
- *Chapter 11, Writing Tests for React and Node.js*

4
Hooking Apollo into React

Sequelize makes it easy to access and query our database. Posts, chats, and messages can be saved to our database in a snap. React helps us to view and update our data by building a **user interface (UI)**.

In this chapter, we will introduce Apollo's React client to our frontend to connect it with the backend. We will query, create, and update post data using our frontend.

This chapter will cover the following topics:

- Installing and configuring Apollo Client
- Sending requests with GQL and Apollo's Query component
- Mutating data with Apollo Client
- Implementing chats and messages
- Pagination in React and GraphQL
- Debugging with the Apollo Client Devtools

Technical requirements

The source code for this chapter is available in the following GitHub repository:

`https://github.com/PacktPublishing/Full-Stack-Web-Development-with-GraphQL-and-React-Second-Edition/tree/main/Chapter04`

Installing and configuring Apollo Client

We have tested our GraphQL **application programming interface** (**API**) multiple times during development. We can now start to implement the data layer of our frontend code. In later chapters, we will focus on other tasks, such as authentication and client-side routing. For now, we will aim to use our GraphQL API with our React app.

To start, we must install the React Apollo Client library. Apollo Client is a GraphQL client that offers excellent integration with React and the ability to easily fetch data from our GraphQL API. Furthermore, it handles actions such as caching and subscriptions, to implement real-time communication with your GraphQL backend. Although Apollo Client is named after the Apollo brand, it is not tied to Apollo Server. You can use Apollo Client with any GraphQL API or schema out there, as long as they follow the protocol standards. You will soon see how perfectly the client merges with our React setup.

As always, there are many alternatives out there. You can use any GraphQL client that you wish with the current API that we have built. This openness is the great thing about GraphQL: it uses an open standard for communication. Various libraries implement the GraphQL standard, and you are free to use any of them.

> **Important Note**
>
> The most well-known alternatives are Relay (which is made by Facebook) and `graphql-request` (which is made by the people behind Prisma). All of these are great libraries that you are free to use. Personally, I mostly rely on Apollo, but Relay is highly recommended as well. You can find a long list of packages related to the GraphQL ecosystem at `https://github.com/chentsulin/awesome-graphql`.

In addition to special client libraries, you could also just use a plain `fetch` method or `XMLHttpRequest` requests. The disadvantage is that you need to implement caching, write `request` objects, and integrate the `request` method into your application on your own. I do not recommend doing this because it takes a lot of time and you want to put that time into your business, not into implementing existing functionalities.

Installing Apollo Client

We use npm to install our client dependencies, as follows:

```
npm install --save @apollo/client graphql
```

We need to install the following two packages to get the GraphQL client running:

- @apollo/client is the wrapping package for all of the packages that we installed. Apollo Client relies on all the other packages.

- graphql is a reference implementation for GraphQL and provides logic to parse GraphQL queries.

You will see how these packages work together in this section.

To get started with the manual setup of Apollo Client, create a new folder and file for the client, as follows:

```
mkdir src/client/apollo
touch src/client/apollo/index.js
```

We will set up Apollo Client in this index.js file. Our first setup will represent the most basic configuration to get a working GraphQL client.

> **Tip**
> The code that follows was taken from the official Apollo documentation. Generally, I recommend reading through the Apollo documentation as it is very well written. You can find this at https://www.apollographql.com/docs/react/essentials/get-started.html.

Just insert the following code:

```
import { ApolloClient, InMemoryCache, from, HttpLink } from '@apollo/client';
import { onError } from "@apollo/client/link/error";

const client = new ApolloClient({
  link: from([
    onError(({ graphQLErrors, networkError }) => {
      if (graphQLErrors) {
        graphQLErrors.map(({ message, locations, path }) =>
        console.log('[GraphQL error]: Message: ${message},
```

```
        Location:
        ${locations}, Path: ${path}'));
        if (networkError) {
          console.log('[Network error]: ${networkError}');
        }
      }
    }),
    new HttpLink({
      uri: 'http://localhost:8000/graphql',
    }),
  ]),
  cache: new InMemoryCache(),
});

export default client;
```

The preceding code uses all of the new packages, apart from `react-apollo`. Let's break down the code, as follows:

- First, at the top of the file, we imported all required functions and classes from the `@apollo/client` package.

- We instantiated `ApolloClient`. For this to work, we passed some parameters, which are the `link` and `cache` properties.

- The `link` property is filled by the `from` command. This function walks through an array of links and initializes each of them, one by one. The links are described further here:

  a. The first link is the error link. It accepts a function that tells Apollo what should be done if an error occurs.

  b. The second link is the **HyperText Transfer Protocol (HTTP)** link for Apollo. You have to offer a **Uniform Resource Identifier (URI)**, under which our Apollo or GraphQL server is reachable. Apollo Client sends all requests to this URI. Notably, the order of execution is the same as the array that we just created.

- The `cache` property takes an implementation for caching. One implementation can be the default package, `InMemoryCache`, or a different cache.

> **Important Note**
> There are many more properties that our links can understand (especially the HTTP link). They feature a lot of different customization options, which we will look at later. You can also find them in the official documentation, at `https://www.apollographql.com/docs/react/`.

In the preceding code snippet, we exported the initialized Apollo Client using the `export default client` line. We are now able to use it in our React app.

The basic setup to send GraphQL requests using Apollo Client is finished. In the next section, we will send our first GraphQL request through Apollo Client.

Testing Apollo Client

Before inserting the GraphQL client directly into our React application tree, we should test it. We will write some temporary code to send our first GraphQL query. After testing our GraphQL client, we will remove the code again. Follow these next steps:

1. Import the package at the top of the Apollo Client setup, as follows:

    ```
    import { gql } from '@apollo/client';
    ```

2. Then, add the following code before the client is exported:

    ```
    client.query({
      query: gql'
        {
          posts {
            id
            text
            user {
              avatar
              username
            }
          }
        }'
    }).then(result => console.log(result));
    ```

The preceding code is almost the same as the example from the Apollo documentation, but I have replaced their query with one that matches our backend.

Here, we used the `gql` tool from Apollo Client to parse a **template literal**. A template literal is just a multiline string surrounded by two grace accents. The `gql` command parses this literal to an **abstract syntax tree** (**AST**). ASTs are the first step of GraphQL; they are used to validate deeply nested objects, the schema, and the query. The client sends our query after the parsing has been completed.

> **Tip**
>
> If you want to know more about ASTs, the people at *Contentful* wrote a great article about what ASTs mean to GraphQL, at `https://www.contentful.com/blog/2018/07/04/graphql-abstract-syntax-tree-new-schema/`.

To test the preceding code, we should start the server and the frontend. One option is to build the frontend now, and then start the server. In this case, the **Uniform Resource Locator** (**URL**) to browse the frontend would be `http://localhost:8000`. A better option would be to spawn the server with `npm run server` and then open a second terminal. Then, you can start the `webpack` development server by executing `npm run client`. A new browser tab should open automatically.

However, we have forgotten something: the client is set up in our new file, but it is not yet used anywhere. Import it in the `index.js` root file of our client React app, below the import of the `App` component, as follows:

```
import client from './apollo';
```

The browser should be reloaded, and the query sent. You should be able to see a new log inside the console of the developer tools of your browser.

The output should look like this:

```
▼ {data: {…}, loading: false, networkStatus: 7}
  ▼ data:
    ▼ posts: Array(2)
      ▼ 0:
          id: 1
          text: "Lorem ipsum 1"
        ▼ user:
            avatar: "/uploads/avatar1.png"
            username: "TestUser"
            __typename: "User"
          ▶ __proto__: Object
          __typename: "Post"
        ▶ __proto__: Object
      ▼ 1:
          id: 2
          text: "Lorem ipsum 2"
        ▼ user:
            avatar: "/uploads/avatar2.png"
            username: "TestUser2"
            __typename: "User"
          ▶ __proto__: Object
          __typename: "Post"
        ▶ __proto__: Object
        length: 2
      ▶ __proto__: Array(0)
    ▶ __proto__: Object
    loading: false
    networkStatus: 7
  ▶ __proto__: Object
```

Figure 4.1 – Manual client response

The data object looks much like the response that we received when sending requests through Postman, except that it now has some new properties: loading and networkStatus. Each of these stands for a specific status, as follows:

- loading, as you might expect, indicates whether the query is still running or has already finished.

- networkStatus goes beyond this and gives you the exact status of what happened. For example, the number 7 indicates that there are no running queries that produce errors. The number 8 means that there has been an error. You can look up the other numbers in the official GitHub repository, at https://github.com/apollographql/apollo-client/blob/main/src/core/networkStatus.ts.

Now that we have verified that the query has run successfully, we can connect Apollo Client to the React **Document Object Model (DOM)**. Please remove the temporary code that we wrote in this section before continuing. This includes everything except the `import` statement in the `App.js` file.

Binding Apollo Client to React

We have tested Apollo Client and have confirmed that it works. However, React does not yet have access to it. Since Apollo Client is going to be used everywhere in our application, we can set it up in our root `index.js` file, as follows:

```
import React from 'react';
import ReactDOM from 'react-dom';
import { ApolloProvider } from '@apollo/client/react';
import App from './App';
import client from './apollo';


ReactDOM.render(
  <ApolloProvider client={client}>
    <App />
  </ApolloProvider>, document.getElementById('root')
);
```

As we mentioned in *Chapter 1*, *Preparing Your Development Environment*, you should only edit this file when the whole application needs access to the new component. In the preceding code snippet, you can see that we import the last package that we installed at the beginning, from `@apollo/client/react`. The `ApolloProvider` component that we extracted from it is the first layer of our React application. It surrounds the `App` component, passing the Apollo Client that we wrote to the next level. To do this, we pass `client` to the provider as a property. Every underlying React component can now access Apollo Client.

We should be now able to send GraphQL requests from our React app.

Using Apollo Client in React

Apollo Client gives us everything that we need to send requests from our React components. We have already tested that the client works. Before moving on, we should clean up our file structure, to make it easier for us later in the development process. Our frontend is still displaying posts that come from static demo data. The first step is to move over to Apollo Client and fetch the data from our GraphQL API.

Follow these instructions to connect your first React component with Apollo Client:

1. Clone the `App.js` file to another file, called `Feed.js`.

2. Remove all parts where React `Helmet` is used, remove the **Cascading Style Sheets**
    (**CSS**) import, and rename the function `Feed` instead of `App`.

3. From the `App.js` file, remove all of the parts that we have left in the `Feed`
    component.

4. Furthermore, we must render the `Feed` component inside of the `App` component.
    It should look like this:

```
import React from 'react';
import { Helmet } from 'react-helmet';
import Feed from './Feed';
import '../../assets/css/style.css';

const App = () => {
  return (
    <div className="container">
      <Helmet>
        <title>Graphbook - Feed</title>
        <meta name="description" content="Newsfeed of
          all your friends on Graphbook" />
      </Helmet>
      <Feed />
    </div>
  )
}

export default App
```

The corresponding `Feed` component should only include the parts where the news feed
is rendered.

We imported the `Feed` component and inserted it inside of the `return` statement of
our `App` component so that it is rendered. The next chapter focuses on reusable React
components and how to write well-structured React code. Now, let's take a look at why
we split our `App` function into two separate files.

Querying in React with Apollo Client

Apollo Client offers one primary way to request data from a GraphQL API. The useQuery function offered by the @apollo/client package provides the ability to request data via React Hooks within a functional React component. Beyond that, you can still rely on the plain client.query function for class-based components if required. There had been multiple ways to do this before, which were deprecated when the new version of Apollo Client was released. Before, you were able to use a **higher-order component** (**HOC**) or use the Query component of Apollo, which is a special React component. Both approaches still exist but are deprecated, so it is not recommended to use them anymore. This is the reason why those approaches will not be explained in this book.

Apollo useQuery Hook

The newest version of Apollo Client comes with the useQuery Hook. You just need to pass the GraphQL query string to the useQuery Hook and it will return you an object that includes data, error, and loading properties that you can use to render your UI.

The actual way of implementing this useQuery Hook is very straightforward. Just follow these instructions:

1. Remove the demo posts from the top of the Feed.js file.

2. Remove the useState(initialPosts) line so that we can query the posts instead.

3. Import the gql function and the useQuery Hook from Apollo and parse the query, as follows:

```
import { gql, useQuery } from '@apollo/client';

const GET_POSTS = gql'{
  posts {
    id
    text
    user {
      avatar
      username
    }
  }
}';
```

4. Execute the useQuery Hook within the Feed function at the top, as follows:

```
const { loading, error, data } = useQuery(GET_POSTS);
```

5. Before the actual return statement, add the following two statements, which will render loading and error messages if there have been any:

```
if (loading) return 'Loading...';
if (error) return 'Error! ${error.message}';
```

6. Beneath these statements and before the last return statement, add this line of code:

```
const { posts } = data;
```

This will make the posts property accessible from the data returned by the useQuery function if it is no longer loading and there is no error.

Note that the function is now way cleaner because we only loop over the posts property and return the markup.

In comparison with the older approaches, the useQuery Hook is easy to understand and also allows us to write readable and understandable code.

The rendered output should look like that shown in *Chapter 1*, *Preparing Your Development Environment*. The form to create a new post is not working at the moment because of our changes; let's fix this in the next section.

Mutating data with Apollo Client

We have replaced the way we get data in our client. The next step is to switch the way in which we create new posts, too. Before Apollo Client, we had to add new fake posts to the array of demo posts manually, within the memory of the browser. Now, everything in our text area is sent with the addPost mutation to our GraphQL API, through Apollo Client.

As with the GraphQL queries, there is a useMutation Hook that you can use to send a mutation against our GraphQL API. Before, there was also an HOC method and a separate Mutation component, which have been deprecated as well. They still exist for backward compatibility, but we will not cover them in this book.

Apollo useMutation Hook

The newest version of Apollo Client comes with the useMutation Hook. The method works equally to the useQuery Hook—you just need to pass the parsed mutation string to it. In response to that, the useMutation Hook will return a function equally named to the mutation, which you can use to trigger those GraphQL requests.

Follow these instructions to implement the useMutation Hook and start using it:

1. Import the useMutation Hook from the @apollo/client package, like this:

    ```
    import { gql, useQuery, useMutation } from '@apollo/
    client';
    ```

2. Parse the addPost mutation string below the getPost query with the gql function, like this:

    ```
    const ADD_POST = gql'
      mutation addPost($post : PostInput!) {
        addPost(post : $post) {
          id
          text
          user {
            username
            avatar
          }
        }
      }
    ';
    ```

3. Inside the Feed component, add the following line of code to get the addPost function, which you can use in the Feed component wherever you want:

    ```
    const [addPost] = useMutation(ADD_POST);
    ```

4. Now that we have got the addPost function, we can start making use of it. Just update the handleSubmit function, as follows:

    ```
    const handleSubmit = (event) => {
      event.preventDefault();
      addPost({ variables: { post: { text: postContent } }
    });
    ```

```
    setPostContent('');
  };
```

As you can see, we completely got rid of the `newPost` object and just send the post's text. Our GraphQL API will create an **identifier (ID)** on insert to the database. As mentioned in *Chapter 3, Connecting to the Database*, we statically add the first user as the author of the post.

You can try to add a new post through the frontend but you won't be able to see it immediately. The form will be empty, but the new post will not be shown. This happens because the current state (or cache) of our component has not yet received the new post. The easiest way to test that everything has worked is to refresh the browser.

Of course, this is not the way that it should work. After the mutation has been sent, the new post should be directly visible in the feed. We will fix this now.

Updating the UI with Apollo Client

After running the `addPost` mutation, the request goes through to the server and saves the new post in our database without any problems. However, we still cannot see the changes take effect in the frontend immediately.

There are two different ways to update the UI after a mutation, as follows:

- **Refetching the dataset**: This is easy to implement but it refetches all of the data, which is inefficient.

- **Updating the cache according to the inserted data**: This is harder to understand and implement, but it attaches the new data to the cache of Apollo Client, so no refetching is needed.

We use these solutions in different scenarios. Let's take a look at some examples. Refetching makes sense if further logic is implemented on the server that is hidden from the client when requesting a list of items and is not applied when inserting only one item. In these cases, the client cannot simulate the state of the typical response of a server.

Updating the cache, however, makes sense when adding or updating items in a list, such as our post feed. The client can insert the new post at the top of the feed.

We will start by simply refetching requests, and then we'll go over the cache update implementation. The following sections (and chapters) will assume that you are not using the HOC method.

Refetching queries

As mentioned previously, this is the easiest method to update your UI. The only step is to set an array of queries to be refetched. The useMutation function should look like this:

```
const [addPost] = useMutation(ADD_POST, {
  refetchQueries: [{query:GET_POSTS}]
});
```

Each object that you enter in the refetchQueries array needs a query property. Each component relying on one of those requests is rerendered when the response for its associated query arrives. It also includes components that are not inside of the Feed component. All components using the post's GET_POSTS query are rerendered.

You can also provide more fields to each query, such as variables to send parameters with the refetch request. Submitting the form resends the query, and you can see the new post directly in the feed. Refetching also reloads the posts that are already showing, which is unnecessary.

Now, let's take a look at how we can do this more efficiently.

Updating the Apollo Client cache

We want to explicitly add only the new post to the cache of Apollo Client. Using the cache helps us to save data by not refetching the complete feed or rerendering the complete list. To update the cache, you should remove the refetchQueries property.

There are technically at least two ways to update the cache on response of the mutation request. The first one is pretty straightforward and simple. You can then introduce a new property, called update, as shown in the following code snippet:

```
const [addPost] = useMutation(ADD_POST, {
  update(cache, { data: { addPost } }) {
    const data = cache.readQuery({ query: GET_POSTS });
    const newData = { posts: [addPost, ...data.posts]};
    cache.writeQuery({ query: GET_POSTS, data: newData });
  }
});
```

The new property runs when the GraphQL addPost mutation has finished. The first parameter that it receives is the cache parameter of Apollo Client, in which the whole cache is saved. The second parameter is the returned response of our GraphQL API.

Updating the cache works like this:

1. Use the `cache.readQuery` function by passing `query` as a parameter. It reads the data, which has been saved for this specific query inside of the cache. The `data` variable holds all of the posts that we have in our feed.

2. Now that we have all of the posts in an array, we can add the missing post. Make sure that you know whether you need to prepend or append an item. In our example, we create a `newData` object with a `posts` array inside it that consists of the newly added post at the top of our list and a destructured list of the old posts.

3. We need to save the changes back to the cache. The `cache.writeQuery` function accepts the `query` parameter that we used to send the request. This `query` parameter is used to update the saved data in our cache. The second parameter is the data that should be saved.

4. When the cache has been updated, our UI reactively renders the changes.

In reality, you can do whatever you want in the `update` function, but we only use it to update the Apollo Client store.

The second way looks a bit more complicated but represents the way that is shown in the official documentation. The `update` function looks a bit more complex but comes with a small improvement of performance. It is your choice which one you like more.

Just replace the `update` function of the `useMutation` Hook, like this:

```
update(cache, { data: { addPost } }) {
  cache.modify({
    fields: {
      posts(existingPosts = []) {
        const newPostRef = cache.writeFragment({
          data: addPost,
          fragment: gql'
            fragment NewPost on Post {
              id
              type
            }
          '
        });
        return [newPostRef, ...existingPosts];
```

```
      }
    }
  });
}
```

What the preceding code does is use the `cache.modify` function, which allows a more precise update than the one we did before. Instead of updating the whole GET_POSTS query within the Apollo Client, we are using the `cache.writeFragment` method of the `cache` object to update the cache and the UI only with the new post. This will improve the performance of our components, especially as the component logic grows.

In the next section, we will be a bit more optimistic about the response of our server and will add the item before the request's response successfully arrives.

Optimistic UI

Apollo provides a great feature of being able to update the UI in an optimistic manner. An optimistic manner means that Apollo adds the new data or post to the storage before the request has finished. The advantage is that the user can see the new result, instead of waiting for the response of the server. This solution makes the application feel faster and more responsive.

This section expects the `update` function of the `Mutation` component to already be implemented. Otherwise, this UI feature will not work. We need to add the `optimisticResponse` property to our mutation next to the `update` property of the `useMutation` configuration, as follows:

```
optimisticResponse: {
  __typename: "mutation",
  addPost: {
    __typename: "Post",
    text: postContent,
    id: -1,
    user: {
      __typename: "User",
      username: "Loading...",
      avatar: "/public/loading.gif"
    }
  }
}
```

The optimisticResponse property can be anything from a function to a simple object. The return value, however, needs to be a GraphQL response object. What you see here is an addPost object that looks as though our GraphQL API could return it, if our request is successful. You need to fill in the __typename fields, according to the GraphQL schema that you are using. That is why the Post and User type names are inside of this fake object.

Technically, you can also add the optimisticResponse property on the actual call to addPost next to the variables property, but I think this is nothing that we need to pass with every call to this function but that actually should be set globally to the useMutation Hook.

The id property of the optimistic response is set to -1. React expects that every component in a loop gets a unique key property. We usually use the id property of a post as the key value. -1 is never used by any other post because MySQL starts counting at 1. Another advantage is that we can use this id property to set a special class to the post item in our list.

Furthermore, the username and the user's avatar are set to loading. That is because we don't have built-in authentication. React and Apollo do not have a user associated with the current session, so we cannot enter the user's data into the optimisticResponse property. We fix this once the authentication is ready. This is an excellent example of how to handle a situation in which you do not have all of the data until you receive a response from the server.

To set a particular class on the list item, we conditionally set the correct className property in our map loop. Insert the following code into the return statement:

```
{posts.map((post, i) =>
  <div key={post.id} className={'post ' + (post.id < 0 ?
    'optimistic': '')}>
    <div className="header">
      <img src={post.user.avatar} />
      <h2>{post.user.username}</h2>
    </div>
    <p className="content">
      {post.text}
    </p>
  </div>
)}
```

An example CSS style for this might look like this:

```
.optimistic {
  -webkit-animation: scale-up 0.4s cubic-bezier(0.390,
    0.575, 0.565, 1.000) both;
  animation: scale-up 0.4s cubic-bezier(0.390, 0.575,
    0.565, 1.000) both;
}


@-webkit-keyframes scale-up {
  0% {
    -webkit-transform: scale(0.5);
    transform: scale(0.5);
  }
  100% {
    -webkit-transform: scale(1);
    transform: scale(1);
  }
}

@keyframes scale-up {
  0% {
    -webkit-transform: scale(0.5);
    transform: scale(0.5);
  }
  100% {
    -webkit-transform: scale(1);
    transform: scale(1);
  }
}
```

CSS animations make your applications more modern and flexible. If you experience issues when viewing these in your browser, you may need to check whether your browser supports them.

You can see the result in the following screenshot:

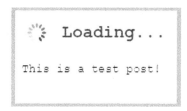

Figure 4.2 – Loading optimistic response

The loading spinner and the username are removed once the response arrives from our API, and the update function is executed again with the real data. You do not need to take care of removing the loading post yourself as this is done by Apollo automatically. Any spinner component from an npm package or GIF file can be used where I have inserted the loading animation. The file that I am using needs to be saved under the public folder, with the name loading.gif, so that it can be used through the CSS we added in the preceding code.

Everything is now set up for sending new posts. The UI responds immediately and shows you the new post.

However, what about new posts from your friends and colleagues? Currently, you need to reload the page to see them, which is not very intuitive. At the moment, we only add the posts that we send on our own but do not receive any information about new posts from other people. I will show you the quickest way to handle this in the following section.

Polling with Apollo Client

Polling is nothing more than rerunning a request after a specified interval. This procedure is the simplest way to implement updates for our news feed. However, multiple issues are associated with polling, as follows:

- It is inefficient to send requests without knowing whether there is any new data. The browser might send dozens of requests without ever receiving a new post.

- If we directly send the initial request again, we will get all of the posts, including those that we are already showing to the user.

- When sending requests, the server needs to query the database and calculate everything. Unnecessary requests cost money and time.

There are some use cases in which polling makes sense. One example is a real-time graph, in which every axis tick is displayed to the user, whether there is data or not. You do not need to use an interrupt-based solution, since you want to show everything. Despite the issues that come with polling, let's quickly run through how it works. All you need to do is fill in the `pollInterval` property in the configuration for the `useQuery` Hook, as follows:

```
const { loading, error, data } = useQuery(GET_POSTS, {
pollInterval: 5000 });
```

The request is resent every 5 seconds (5,000 **milliseconds**, or **ms**).

As you might expect, there are other ways to implement real-time updates to your UI. One approach is to use **server-sent events**. A server-sent event is, as the name suggests, an event that is sent by the server to the client. The client needs to establish a connection to the server, but then the server can send messages to the client, in one direction. Another method is to use **WebSockets**, which allow for bidirectional communication between the server and the client. The most common method in relation to GraphQL, however, is to use **Apollo Subscriptions**. They are based on WebSockets and work perfectly with GraphQL. I will show you how Apollo Subscriptions work in *Chapter 10, Real-Time Subscriptions*.

Let's continue and integrate the rest of our GraphQL API.

Implementing chats and messages

In the previous chapter, we programmed a pretty dynamic way of creating chats and messages with your friends and colleagues, either one-on-one or in a group. There are some things that we have not discussed yet, such as authentication, real-time subscriptions, and friend relationships. First, however, we are going to work on our new skills, using React with Apollo Client to send GraphQL requests. It is a complicated task, so let's get started.

Fetching and displaying chats

Our news feed is working as we expected. Now, we also want to cover chats. As with our feed, we need to query for every chat that the current user (or, in our case, the first user) is associated with.

The initial step is to get the rendering working with some demo chats. Instead of writing the data on our own, as we did in the first chapter, we can now execute the `chats` query. Then, we can copy the result into the new file as static demo data, before executing the actual `useQuery` Hook.

Let's get started, as follows:

1. Send the GraphQL query. The best options involve the Apollo Client Devtools if
    you already know how they work. Otherwise, you can rely on Postman, as you did
    previously. The code is illustrated in the following snippet:

    ```
    query {
      chats {
        id
        users {
          avatar
          username
        }
      }
    }
    ```

    The request looks a bit different from the one we tested with Postman. The chat
    panel that we are going to build only needs specific data. We do not need to render
    any messages inside of this panel, so we don't need to request them. A complete
    chat panel only requires the chat itself, the ID, the usernames, and the avatars.
    Later, we will retrieve all of the messages, too, when viewing a single chat.

    Next, create a new file called Chats.js, next to the Feed.js file.

    Copy the complete chats array from the response over to an array inside of the
    Chats.js file, as follows. Add it to the top of the file:

    ```
    const chats = [{
      "id": 1,
      "users": [{
        "id": 1,
        "avatar": "/uploads/avatar1.png",
        "username": "Test User"
      },
      {
        "id": 2,
        "avatar": "/uploads/avatar2.png",
        "username": "Test User 2"
      }]
    }
    ];
    ```

2. Import React ahead of the `chats` variable. Otherwise, we will not be able to render any React components. Here's the code you'll need to do this:

```
import React, { useState } from 'react';
```

3. Set up the functional React component. I have provided the basic markup here. Just copy it beneath the `chats` variable. I am going to explain the logic of the new component shortly:

```
const usernamesToString = (users)  => {
  const userList = users.slice(1);
  var usernamesString = '';

  for(var i = 0; i < userList.length; i++) {
    usernamesString += userList[i].username;
    if(i - 1 === userList.length) {
      usernamesString += ', ';
    }
  }
  return usernamesString;
}
const shorten = (text) => {
  if (text.length > 12) {
    return text.substring(0, text.length - 9) + '...';
  }

  return text;
}

const Chats = () => {
  return (
    <div className="chats">
      {chats.map((chat, i) =>
        <div key={chat.id} className="chat">
          <div className="header">
            <img src={(chat.users.length > 2 ?
              '/public/group.png' :
                chat.users[1].avatar)} />
```

```
            <div>
              <h2>{shorten(usernamesToString
                (chat.users))}</h2>
            </div>
          </div>
        </div>
      )}
    </div>
  )
}
```

```
export default Chats
```

The component is pretty basic, at the moment. The component maps over all of the chats and returns a new list item for each chat. Each list item has an image that is taken from the second user of the array, since we defined that the first user in the list is the current user, as long as we have not implemented authentication. We use a group icon if there are more than two users. When we have implemented authentication and we know the logged-in user, we can take the specific avatar of the user that we are chatting with.

The title displayed inside of the h2 tag at the top of the chat is the name of the user. For this, I have implemented the usernamesToString method, which loops over all of the usernames and concatenates them into a long string. The result is passed into the shorten function, which removes all of the characters of the string that exceed the size of the maximum-12 characters.

One thing you may notice is that these helper functions are not within the actual component. I personally recommend having helper functions outside of the component as they will be recreated on every render of the component. If the helper function needs the scope of the component, keep it inside, but if they are pure functions just doing transformations here, keep them outside.

4.   Our new component needs some styling. Copy the new CSS to our style.css file.

To save the file size in our CSS file, replace the two .post .header styles to also cover the style of the chats, as follows:

```
.post .header > *, .chats .chat .header > * {
  display: inline-block;
  vertical-align: middle;
```

```
}

.post .header img, .chats .chat .header img {
  width: 50px;
  margin: 5px;
}
```

We must append the following CSS to the bottom of the `style.css` file:

```
.chats {
  background-color: #eee;
  width: 200px;
  height: 100%;
  position: fixed;
  top: 0;
  right: 0;
  border-left: 1px solid #c3c3c3;
}

.chats .chat {
  cursor: pointer;
}

.chats .chat .header > div {
  width: calc(100% - 65px);
  font-size: 16px;
  margin-left: 5px;
}

.chats .chat .header h2, .chats .chat .header span {
  color: #333;
  font-size: 16px;
  margin: 0;
}

.chats .chat .header span {
```

```
    color: #333;
    font-size: 12px;
  }
```

5. To get the code working, we must also import the `Chats` component into our `App.js` file, as follows:

    ```
    import Chats from './Chats';
    ```

6. Render the `Chats` component inside the return statement beneath the `Feed` component inside of the `App.js` file.

    The current code generates the following screenshot:

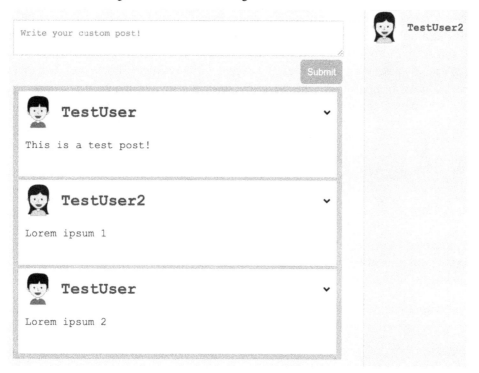

Figure 4.3 – Chats panel

On the right-hand side, you can see the chats panel that we have just implemented. Every chat is listed there as a separate row.

The result isn't bad, but it would be much more helpful to at least have the last message of every chat beneath the username so that you could directly see the last content of your conversations.

Just follow these instructions to get the last message into the chats panel:

1. The easiest way to do this would be to add the messages to our query again, but querying all of the messages for every chat that we want to display in the panel would not make much sense. Instead, we will add a new property to the chat entity, called `lastMessage`. That way, we will only get the newest message. We will add the new field to the GraphQL schema of our chat type, in the backend code, as follows:

```
lastMessage: Message
```

Of course, we must also implement a function that retrieves the `lastMessage` field.

2. Adding our new `resolvers.js` function inside the `Chats` property of the `resolvers` object orders all of the chat messages by ID and takes the first one. By definition, this should be the latest message in our chat. We need to resolve the promise on our own and return the first element of the array since we expect to return only one `message` object. If you return the promise directly, you will receive `null` in the response from the server because an array is not a valid response for a single message entity. The code is illustrated in the following snippet:

```
lastMessage(chat, args, context) {
  return chat.getMessages({limit: 1, order: [['id',
    'DESC']]}).then((message) => {return message[0];
  });
},
```

3. You can add the new property to our static demo data inside `Chats.js.` for every array item or rerun the GraphQL query and copy the response again. The code is illustrated here:

```
"lastMessage": {
  "text": "This is a third test message."
}
```

4. We can render the new message with a simple `span` tag beneath the h2 header of the username. Copy it directly into the `return` statement, inside of our `Chats` component, as follows:

```
<span>{chat?.lastMessage?.text}</span>
```

The result of the preceding changes renders every chat row with the last message inside of the chat. It should now look like this:

TestUser2

This is a second
test...

Figure 4.4 – Last message

Since everything is displayed correctly from our test data, we can introduce the useQuery Hook in order to fetch all of the data from our GraphQL API. We can remove the chats array. Then, we will import all of the dependencies and parse the GraphQL query, as in the following code snippet:

```
import { gql, useQuery } from '@apollo/client';

const GET_CHATS = gql'{
  chats {
    id
    users {
      id
      avatar
      username
    }
    lastMessage {
      text
    }
  }
}';
```

To make use of the preceding parsed GraphQL query we will execute the useQuery Hook in our functional component, as follows:

```
const { loading, error, data} = useQuery(GET_CHATS);

if (loading) return <div className="chats"><p>Loading...</p></div>;
```

```
if (error) return <div className="chats"><p>{error.message}</
p></div>;

const { chats } = data;
```

When you have added the preceding lines of code to the beginning of the Chats function, it will then make use of the chats array returned within the GraphQL response. Before doing so it will, of course, check if the request is still loading or if there has been an error.

We render the loading and error state within the div tag with the chats class so that the messages are wrapped within the gray panel.

You should have run the addChat mutation from the previous chapter through Postman. Otherwise, there will be no chats to query for, and the panel will be empty. You have to also execute this mutation for any following chapter because we are not going to implement a special button for this functionality. The reason is that the logic behind it does not provide further knowledge about React or Apollo as it is just done by executing the addChat mutation at the correct location within Graphbook.

Next, we want to display chat messages after opening a specific chat.

Fetching and displaying messages

First, we have to store the chats that were opened by a click from the user. Every chat is displayed in a separate, small chat window, like on Facebook. Add a new state variable to save the IDs of all of the opened chats to the Chats component, as follows:

```
const [openChats, setOpenChats] = useState([]);
```

To let our component insert or remove something from the array of open chats, we will add the new openChat and closeChat functions, as follows:

```
const openChat = (id) => {
  var openChatsTemp = openChats.slice();

  if(openChatsTemp.indexOf(id) === -1) {
    if(openChatsTemp.length > 2) {
      openChatsTemp = openChatsTemp.slice(1);
    }
    openChatsTemp.push(id);
```

```
    }

    setOpenChats(openChatsTemp);
}

const closeChat = (id) => {
    var openChatsTemp = openChats.slice();

    const index = openChatsTemp.indexOf(id);
    openChatsTemp.splice(index,1),

    setOpenChats(openChatsTemp);
}
```

When a chat is clicked on, we will first check that it is not already open, by searching the ID using the indexOf function inside of the openChats array.

Every time a new chat is opened, we will check whether there are three or more chats. If that is the case, we will remove the first opened chat from the array and exchange it with the new one by appending it to the array with the push function. We will only save the chat IDs, not the whole **JavaScript Object Notation (JSON)** object.

For the closeChat function, we just revert this by removing the ID from the openChats array.

The last step is to bind the onClick event to our component. In the map function, we can replace the wrapping div tag with the following line of code:

```
<div key={"chat" + chat.id} className="chat" onClick={() =>
openChat(chat.id)}>
```

Here, we use onClick to call the openChat function, with the chat ID as the only parameter. At this point, the new function is already working but the updated state isn't used. Let's take care of that, as follows:

1. Add a surrounding wrapper div tag to the div tag with the chats class, as follows:

    ```
    <div className="wrapper">
    ```

2. In order to not mess up, the complete code we have written will introduce our first child component. To do so, create a file called Chat.js next to the Chats.js file.

3. In this new file, import React and Apollo and parse the GraphQL query to get all the chat messages that were just opened, as follows:

```
import React from 'react';
import { gql, useQuery } from '@apollo/client';


const GET_CHAT = gql'
  query chat($chatId: Int!) {
    chat(chatId: $chatId) {
      id
      users {
        id
        avatar
        username
      }
      messages {
        id
        text
        user {
          id
        }
      }
    }
  }
';
```

As you can see in the preceding code snippet, we are passing the chat ID as a parameter to the GraphQL query.

4. The actual component will then make use of the parsed query to get all the messages and render them inside a small container. The component should be added, as follows:

```
const Chat = (props) => {
  const { chatId, closeChat } = props;
  const { loading, error, data } = useQuery(GET_CHAT, {
    variables: { chatId }});
```

```
  if (loading) return <div className="chatWindow">
    <p>Loading...</p></div>;
  if (error) return <div className="chatWindow">
    <p>{error.message}</p></div>;

  const { chat } = data;

  return (
    <div className="chatWindow">
      <div className="header">
        <span>{chat.users[1].username}</span>
        <button onClick={() => closeChat(chatId)}
          className="close">X</button>
      </div>
      <div className="messages">
        {chat.messages.map((message, j) =>
          <div key={'message' + message.id}
            className={'message ' + (message.user.id >
              1 ? 'left' : 'right')}>
            {message.text}
          </div>
        )}
      </div>
    </div>
  )
}

export default Chat
```

We execute the `useQuery` Hook to send the GraphQL request. We pass the
`chatId` property from the `props` property so that the chat ID must be passed to
this child component from the parent component. We also extract the `closeChat`
function to call it from the child component, as the actual close button is within the
chat container and not within the parent component.

Once the request arrives, we check again if the request is loading or has an error before actually rendering the complete chat. Then, we render a `div` tag with the `chatWindow` class name, in which all messages are displayed. Again, we are using the user ID to fake the class name of the messages. We will replace it when we get authentication running.

5. As we have prepared the child component, we only need to add one line to the `Chats.js` file to import it, as follows:

    ```
    import Chat from './Chat';
    ```

6. Then, to make use of our new `Chat` component, just add these three lines of code inside the `div` tag with the `wrapper` class:

    ```
    <div className="openChats">
        {openChats.map((chatId, i) => <Chat chatId={chatId}
        key={"chatWindow" + chatId} closeChat={closeChat} /> )}
    </div>
    ```

    For each item in the `openChats` array, we will render the `Chat` component, which will then pass the `chatId` property and the `closeChat` function. The child component will then fetch the chat data on its own by the passed `chatId` property.

7. The last thing missing is some styling. The CSS file is pretty big. Every message from the other users should be displayed on the left and our own messages on the right, to differentiate them. Insert the CSS code directly from the GitHub repository to save some time: `https://github.com/PacktPublishing/Full-Stack-Web-Development-with-GraphQL-and-React-2nd-Edition/blob/main/Chapter04/assets/css/style.css`.

Take a look at the following screenshot:

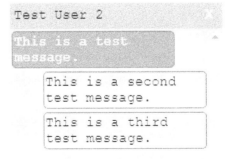

Figure 4.5 – Chat window

We have forgotten something important. We can see all of the messages from our chat, but we are not able to add new messages, which is essential. Let's take a look at how to implement a chat message form in the next section.

Sending messages through mutations

The `addMessage` mutation already exists in our backend, so we can add it to our `Chat` component. To implement this feature completely, follow these instructions:

1. Before adding it straightaway to the frontend too, we need to change the `import` statements so that we also have the `useMutation` and `useState` functions, as follows:

```
import React, { useState } from 'react';
import { gql, useQuery, useMutation } from '@apollo/
client';
```

2. Then, parse the mutation at the top, next to the other requests, like this:

```
const ADD_MESSAGE = gql'
  mutation addMessage($message : MessageInput!) {
    addMessage(message : $message) {
      id
      text
      user {
        id
      }
    }
  }
';
```

3. For now, we will keep it simple and just add the text input to the `Chat` component directly, but we will take a look at a better way to do this in *Chapter 5, Reusable React Components and React Hooks*.

   Now, we need to create a state variable where we save the current value of our new text input that we still need to create. We need to execute the `useMutation` Hook to send the GraphQL request to create a new chat message. Just add the following code for that:

```
const [text, setText] = useState('');
const [addMessage] = useMutation(ADD_MESSAGE, {
  update(cache, { data: { addMessage } }) {
```

```
        cache.modify({
          id: cache.identify(data.chat),
          fields: {
            messages(existingMessages = []) {
              const newMessageRef = cache.writeFragment({
                data: addMessage,
                fragment: gql'
                  fragment NewMessage on Chat {
                    id
                    type
                  }
                '
              });
              return [...existingMessages, newMessageRef];
            }
          }
        });
      }
    });

  const handleKeyPress = (event) => {
    if (event.key === 'Enter' && text.length) {
      addMessage({ variables: { message: { text, chatId
      } } }).then(() => {
        setText('');
      });
    }
  }
```

The state variable and mutation function look familiar, as you already know them. One special thing we do for the useMutation Hook is to again provide an update function to efficiently update the Apollo Client cache with the newest data. To do so, we must provide an id property to the cache.modify function. The reason we need to do that is that we want to update the messages array of one specific chat, but there could be multiple within our cache. To update the correct chat in the message, we use the cache.identify function and provide the current chat object, and it will automatically detect which chat to update.

The `handleKeyPress` function will handle the submission of the text input to trigger a mutation request.

4. We must insert the markup needed to render a fully functional input. Put the input below the messages list, inside of the chat window. The onChange property executes while typing and will update the state of the component with the value of the input. Insert the following code:

```
<div className="input">
  <input type="text" value={text} onChange={ (e) =>
    setText(e.target.value) }
    onKeyPress={handleKeyPress}/>
</div>
```

We use the `onKeyPress` event to handle *Enter* key hits so that we can send the chat message.

5. Let's quickly add some CSS to our `style.css` file to make the input field look good, as follows:

```
.chatWindow .input input {
  width: calc(100% - 4px);
  border: none;
  padding: 2px;
}
.chatWindow .input input:focus {
  outline: none;
}
```

The following screenshot shows the chat window, with a new message inserted through the chat window input:

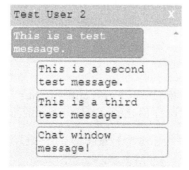

Figure 4.6 – Messaging in the chat window

There are many features that we have not implemented and that we won't cover in this book—for example, it would make sense to have the username next to the chat message if it is a group chat, to show the avatar next to the message, and to update the `lastMessage` field in the chats panel once a new message is sent. The workload required to achieve a fully-fledged social network, such as Facebook, is impossible to cover in this book, but you are going to learn all of the required techniques, tools, and tactics so that you can approach this on your own. The next important feature that we are going to cover is pagination.

Pagination in React and GraphQL

By **pagination**, most of the time, we mean the batch querying of data. Currently, we query for all posts, chats, and messages in our database. If you think about how much data Facebook stores inside one chat with your friends, you will realize that it is unrealistic to fetch all of the messages and data ever shared at once. A better solution is to use pagination. With pagination, we always have a page size, or a limit, of how many items we want to fetch per request. We also have a page or offset number, from which we can start to select data rows.

In this section, we're going to look at how to use pagination with the posts feed, as it is the most straightforward example. In *Chapter 5*, *Reusable React Components and React Hooks*, we will focus on writing efficient and reusable React code. Sequelize offers the pagination feature by default. We can first insert some more demo posts so that we can paginate in batches of 10.

We need to adjust the backend a bit before implementing it on our frontend, as follows:

1. Add a new `RootQuery` to our GraphQL schema, as follows:

    ```
    postsFeed(page: Int, limit: Int): PostFeed
    ```

2. The `PostFeed` type only holds the `posts` field. Later on, in the development of the application, you can return more information, such as the overall count of items, the page count, and so on. The code is illustrated in the following snippet:

    ```
    type PostFeed {
        posts: [Post]
    }
    ```

3. Next, we must implement the `PostFeed` entity in our `resolvers.js` file. Copy the new resolver function over to the `resolvers` file, as follows:

    ```
    postsFeed(root, { page, limit }, context) {
        var skip = 0;
    ```

```
    if(page && limit) {
        skip = page * limit;
    }

    var query = {
        order: [['createdAt', 'DESC']],
        offset: skip,
    };

    if(limit) {
        query.limit = limit;
    }

    return {
        posts: Post.findAll(query)
    };
},
```

We build a simple `query` object that Sequelize understands, which allows us to paginate our posts. The `page` number is multiplied by the `limit` parameter, to skip the calculated number of rows. The `offset` parameter skips the number of rows, and the `limit` parameter stops selecting rows after a specified number (which, in our case, is 10).

Our frontend needs some adjustments to support pagination. Install a new React package with npm, which provides us with an infinite scroll implementation, as follows:

```
npm install react-infinite-scroll-component --save
```

Infinite scrolling is an excellent method to let a user load more content by scrolling to the bottom of the browser window.

You are free to program this on your own, but we are not going to cover that here. Go back to the `Feed.js` file, replace the `GET_POSTS` query, and import the `react-infinite-scroll-component` package with the following code:

```
import InfiniteScroll from 'react-infinite-scroll-component';

const GET_POSTS = gql'
    query postsFeed($page: Int, $limit: Int) {
        postsFeed(page: $page, limit: $limit) {
```

```
        posts {
            id
            text
            user {
                avatar
                username
            }
        }
    }
}
';
```

Since the postsFeed query expects parameters other than the standard query from before, we need to edit our useQuery Hook and also introduce two new state variables. The changed lines are shown here:

```
const [hasMore, setHasMore] = useState(true);
const [page, setPage] = useState(0);
const { loading, error, data, fetchMore } = useQuery(GET_POSTS,
{ pollInterval: 5000, variables: { page: 0, limit: 10 } });
```

In the preceding code snippet, we extract the fetchMore function from the useQuery Hook, which is used to run the pagination request to load more post items. We also create a hasMore state variable that will identify if there is more data to load from the GraphQL API, and the page variable will save the current page—or, to be exact, the number of pages we already scrolled.

According to the new data structure defined in our GraphQL schema, we extract the posts array from the postsFeed object. You can do that by replacing the code with these two lines:

```
const { postsFeed } = data;
const { posts } = postsFeed;
```

Replace the markup of the div tag of our current feed to make use of our new infinite scroll package, as follows:

```
<div className="feed">
  <InfiniteScroll
    dataLength={posts.length}
    next={() => loadMore(fetchMore)}
```

```
    hasMore={hasMore}
    loader={<div className="loader" key={"loader"}>
      Loading ...</div>}
  >
    {posts.map((post, i) =>
      <div key={post.id} className={'post ' + (post.id < 0
        ? 'optimistic': '')}>
        <div className="header">
          <img src={post.user.avatar} />
          <h2>{post.user.username}</h2>
        </div>
        <p className="content">{post.text}</p>
      </div>
    )}
  </InfiniteScroll>
</div>
```

The only thing that the infinite scroll package does is run the `loadMore` function provided in the `next` property, as long as `hasMore` is set to `true` and the user scrolls to the bottom of the browser window. When `hasMore` is set to `false`, the event listeners are unbound and no more requests are sent. This behavior is great when no further content is available, as we can stop sending more requests.

We need to implement the `loadMore` function before running the infinite scroller. It relies on the `page` variable that we just configured. The `loadMore` function should look like this:

```
const loadMore = (fetchMore) => {
  const self = this;

  fetchMore({
    variables: {
      page: page + 1,
    },
    updateQuery(previousResult, { fetchMoreResult }) {
      if(!fetchMoreResult.postsFeed.posts.length) {
        setHasMore(false);
        return previousResult;
```

```
    }

    setPage(page + 1);

    const newData = {
      postsFeed: {
        __typename: 'PostFeed',
        posts: [
          ...previousResult.postsFeed.posts,
          ...fetchMoreResult.postsFeed.posts
        ]
      }
    };
    return newData;
  }
  });
}
```

Let's quickly go through the preceding code, as follows:

1. The fetchMore function receives an object as a parameter.

2. We specify the variables field, which is sent with our request, to query the correct page index of our paginated posts.

3. The updateQuery function is defined to implement the logic to add the new data that needs to be included in our news feed. We can check whether any new data is included in the response by looking at the returned array length. If there are no posts, we can set the hasMore state variable to false, which unbinds all scrolling events. Otherwise, we can continue and build a new postsFeed object inside of the newData variable. The posts array is filled by the previous posts query result and the newly fetched posts. In the end, the newData variable is returned and saved in the client's cache.

4. When the updateQuery function is finished, the UI rerenders accordingly.

At this point, your feed is able to load new posts whenever the user visits the bottom of the window. We no longer load all posts at once, but instead, we only get the 10 most recent from our database. Every time you build an application with large lists and many rows, you have to add some kind of pagination, with either infinite scrolling or simple page buttons.

We have now created a new problem. We can submit a new post with the GraphQL mutation if the React Apollo cache is empty, but the `update` function of the `Mutation` component will throw an error. Our new query is stored not only under its name but also under the variables used to send it. To read the data of a specific paginated `posts` request from our client's cache, we must also pass variables, such as the page index. Furthermore, we have a second layer, `postsFeed`, as the parent of the `posts` array. Change the `update` function to get it working again, as follows:

```
postsFeed(existingPostsFeed) {
  const { posts: existingPosts } = existingPostsFeed;
  const newPostRef = cache.writeFragment({
    data: addPost,
    fragment: gql'
      fragment NewPost on Post {
        id
        type
      }
    '
  });
  return {
    ...existingPostsFeed,
    posts: [newPostRef, ...existingPosts]
  };
}
```

We actually just changed the `posts` property to `postsFeed` and updated the function to update the extracted `posts` array.

Complex code such as this requires some useful tools to debug it. Continue reading to learn more about the Apollo Client Devtools.

Debugging with the Apollo Client Devtools

Whenever you write or extend your own application, you have to test, debug, and log different things during development. In *Chapter 1*, *Preparing Your Development Environment*, we looked at the React Developer Tools for Chrome, while in *Chapter 2*, *Setting Up GraphQL with Express.js*, we explored Postman for testing APIs. Now, let's take a look at another tool.

The **Apollo Client Devtools** is another Chrome extension, allowing you to send Apollo requests. While Postman is great in many ways, it does not integrate with our application and does not implement all GraphQL-specific features. The Apollo Client Devtools rely on the Apollo Client that we set up very early on in this chapter.

Every request, either a query or mutation, is sent through the Apollo Client of our application. The developer tools also provide features such as autocomplete, for writing requests. They can show us the schema as it is implemented in our GraphQL API, and we also can view the cache. We will go over all four of the main windows offered by the extension.

Let's take a look at an example here:

Figure 4.7 – The Apollo Client Devtools

The **GraphiQL** window is shown in the preceding screenshot. The three panels in the preceding screenshot are described as follows:

- You can enter a request that you want to send in the left-hand text area. It can be a mutation or query, including the markup for inputs, for example. You can also enter the variables at the bottom.

- When sending a request, the response is shown in the middle panel.

- In the panel on the right, you can find the schema against which you will run requests. You can search through the complete GraphQL schema or manually step into the tree by clicking on the root types. This feature is useful when you forget what a specific field or mutation is called or which parameters it accepts.

In the top bar, you will find the **Prettify** button, which tidies your query so that it is more readable. The **Load from cache** checkbox tries to retrieve any requested data directly from the cache, when possible. By clicking on the **Play** button, you run the query. These are all tools to test our GraphQL requests properly. The **Build** button will give you a small graphical interface to edit your query.

Next, there is the **Queries** window, which is a helpful display. All of the queries that were ever run through the client are listed here, including the query string and variables. If you want to, you can rerun a query by clicking on the button at the top, as illustrated in the following screenshot:

Figure 4.8 – Apollo Queries window

The **Mutations** window is actually the same as the **Queries** window, but for mutations. The list is empty, as long as you have not sent any mutations.

The last window is **Cache**. Here, you are able to see all of the data stored inside the Apollo cache, as illustrated in the following screenshot:

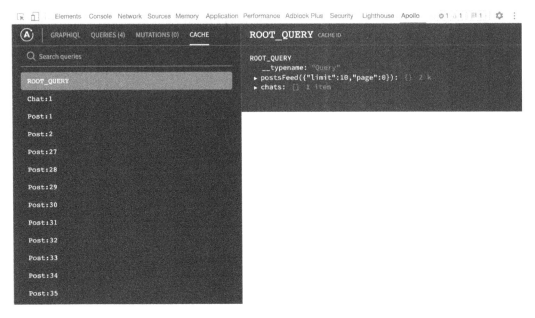

Figure 4.9 – Apollo Cache window

In the left-hand panel, you can search through your data. The right-hand panel shows you the selected object in JSON.

You can also see that I have tested the API a lot, as there are multiple Post objects in the left-hand panel.

> **Resetting the Apollo Cache**
>
> For testing purposes, I submitted multiple posts via a mutation, but I deleted them to make sure that the screenshots were clear. Apollo did not delete the old posts that were deleted in the database, so they are still inside of the cache. You should delete this data when a user logs out of your application so that unauthorized users cannot access it.

That is everything you need to know about the Apollo Client Devtools.

Summary

In this chapter, you learned how to connect your GraphQL API to React. To do this, we used Apollo Client to manage the cache and the state of our components and to update React and the actual DOM of the browser. We looked at how to send queries and mutations against our server. We also covered how to implement pagination with React and Apollo, and how to use the Apollo Client Devtools.

After this chapter, you should be able to integrate the Apollo Client into your React application at any time. Furthermore, you should be able to make use of Apollo in every component of your application and be able to debug it.

The next chapter will cover how to write reusable React components. Up to this point, we have written the code, but we haven't thought about readability or good practices very much. We will address these issues in the next chapter.

5

Reusable React Components and React Hooks

We have done a lot to reach this point in the book, including saving, requesting, inserting, and updating data through the use of Apollo Client, in connection with our GraphQL **application programming interface (API)**. Much of the code that we have written will also have to be reviewed many times. This is especially important because we are building an application so quickly. Everything is working for now, but we have not done a great job here; there are some best practices and tactics that need to be observed in order to write good React applications.

This chapter will cover everything you need to know to write efficient and reusable React components. It will cover the following topics:

- Introducing React patterns
- Structuring our React application
- Extending Graphbook
- Documenting React applications

Technical requirements

The source code for this chapter is available in the following GitHub repository:

`https://github.com/PacktPublishing/Full-Stack-Web-Development-with-GraphQL-and-React-Second-Edition/tree/main/Chapter05`

Introducing React patterns

With any programming language, framework, or library that you use, there are always common tactics that you should follow. They present an understandable, efficient way to write applications.

In *Chapter 4*, *Hooking Apollo into React*, we tackled some patterns, such as rendering arrays, the spread operator, and destructuring objects. Nevertheless, there are some further patterns that you should know about.

We will go over the most commonly used patterns that React offers, as follows:

- Controlled components
- Functional components
- Conditional rendering
- Rendering children

Many (but not all) of the examples here only represent illustrations of what each method looks like. Some of them will not be taken over to our real application code, so if you are not interested in learning the essential aspects of patterns or if you already know most of them, you can skip the examples.

> **Note**
> Beyond the short explanation that I will provide, there is more extensive documentation on this topic. The official React documentation is always a good starting point, but you can find all React patterns, including those that we have already used, at `https://reactpatterns.com/`.

Controlled components

When we wrote our post form to submit new posts or the message inputs inside chat in the previous chapters, we used controlled input by incident. To provide a better understanding, I am going to quickly explain the difference between controlled and uncontrolled components, and when to use each of them.

Let's start with uncontrolled input.

By definition, a component is uncontrolled whenever the value is not set by a property through React, but only saved and taken from the real browser **Document Object Model (DOM)**. The value of an input is then retrieved from a reference to the DOM Node and is not managed and taken from React's component state.

The following code snippet shows the post form where the user will be able to submit new posts. I have excluded the rendering logic for the complete feed, as it is not a part of the pattern that I want to show you:

```javascript
import React, { useState, useRef } from 'react';
import { gql, useQuery, useMutation } from '@apollo/client';

const ADD_POST = gql'
  mutation addPost($post : PostInput!) {
    addPost(post : $post) {
      id
      text
      user {
        username
        avatar
      }
    }
  }
';

const Feed = () => {
  const textArea = useRef(null)
  const [addPost] = useMutation(ADD_POST);

  const handleSubmit = (event) => {
    event.preventDefault();
    addPost({ variables: { post: { text:
      textArea.current.value } } });
  };

  return (
```

```
    <div className="container">
      <div className="postForm">
        <form onSubmit={handleSubmit}>
          <textarea ref={textArea} placeholder="Write your
            custom post!"/>
          <input type="submit" value="Submit" />
        </form>
      </div>
    </div>
  )
}

export default Feed
```

In this example, you can see that we no longer have the useState Hook, since the textarea value is stored within the real DOM Node, and not the application state.

We run the useRef Hook provided by React. It prepares the variable to accept the DOM Node as a property. If you are using class-based React components, you can use the createRef function.

In the return statement for rendering the component, the ref property fills in the reference that we just created with the DOM element.

Accessing the value of the DOM Node works by using the normal JavaScript DOM API. You can see this behavior when sending the submit event of our form. The value is extracted from the textArea.current.value field.

Everything that an uncontrolled component needs is already shown here; there is no more to it. You can compare this approach to our current implementation of the post form. In our implementation, we set up the state, listen for change events, and save and read the value directly from the component state, not from the DOM element.

When using uncontrolled components and working directly with DOM elements, the problem is that you leave the normal React workflow. You are no longer able to handle conditions and, therefore, trigger other events inside of React.

Nevertheless, the DOM reference can make it easier to use third-party plugins that were not written for the React ecosystem. There are thousands of great jQuery plugins, for example. I always recommend using the default approach of a controlled component. For 99% of cases, this works without leaving the React workflow.

> **Tip**
> If you need a deeper understanding of which approach is a better solution
> for your specific case, take a look at `https://goshakkk.name/`
> `controlled-vs-uncontrolled-inputs-react/`.

Functional components

One fundamental and efficient solution for writing well-structured and reusable React
components is the use of functional components. We have already used them throughout
the book at every location.

Before the existence of functional components, there were stateless functions. As you
might expect, stateless functions are functions, not React components. They are not able
to store any states; only properties can be used to pass and render data. Property updates
are directly rerendered inside the stateless functions.

Since newer versions of React, those stateless functions no longer exist as there are React
Hooks that allow us to use state in those functions, which makes them fully functional
components. This is why they are now called functional components.

We have written a lot of code where more functional components can help us provide
more structured and understandable code.

Beginning with the file structure, we will create a new folder for our new components,
as follows:

```
mkdir src/client/components
```

Many parts of our application need to be reworked. Create a new file for our first
functional component, as follows:

```
touch src/client/components/loading.js
```

Currently, we display a dull and boring **Loading...** message when our GraphQL requests
are running. Let's change this by inserting the following code into the `loading.js` file:

```
import React from 'react';

export default ({color, size}) => {
  var style = {
    backgroundColor: '#6ca6fd',
    width: 40,
    height: 40,
```

```
  };

  if(typeof color !== typeof undefined) {
    style.color = color;
  }
  if(typeof size !== typeof undefined) {
    style.width = size;
    style.height = size;
  }


  return <div className="bouncer" style={style}></div>
}
```

In the preceding code snippet, we are using a simple function in **ECMAScript 6 (ES6)** arrow notation. This will actually be a pure stateless function as we do not use any React Hook, as it is just not required for this simple component. In the code, you can see that we are extracting the `color` and `size` fields from the properties that our function receives.

We are building a default `style` object that represents the basic styling for a loading spinner. You can pass the `color` and `size` values separately, to adjust those settings.

Lastly, we are returning a simple `div` tag with the **Cascading Style Sheets (CSS)** style and the `bouncer` class.

What's missing here is the CSS styling. The code should look like this; we'll just add it to our `style.css` file:

```
.bouncer {
  margin: 20px auto;
  border-radius: 100%;
  -webkit-animation: bounce 1.0s infinite ease-in-out;
  animation: bounce 1.0s infinite ease-in-out;
}


@-webkit-keyframes bounce {
  0% {
    -webkit-transform: scale(0)
  }
```

```
    100% {
        -webkit-transform: scale(1.0);
        opacity: 0;
    }
}


@keyframes bounce {
    0% {
        -webkit-transform: scale(0);
        transform: scale(0);
    }
    100% {
        -webkit-transform: scale(1.0);
        transform: scale(1.0);
        opacity: 0;
    }
}
```

As in the previous examples, we use CSS animations to display our loading spinner correctly and to let it animate as pulsating.

We have now finished the functional component. You should place it into the existing code, wherever a loading state exists.

First, import the new loading spinner to the top of your files, as follows:

```
import Loading from './components/loading';
```

You can then render the functional component, as follows:

```
if (loading) return <Loading />;
```

Start the server with npm run server and the frontend with npm run client. You should now see a pulsating blue bubble where you inserted it. I have tested this inside of my posts feed, and it looks pretty good.

The advantage of functional components is that they are minimal and efficient functions, rendering smaller parts of our application. The approach perfectly integrates with React, and we can improve upon the code that we have written.

Conditional rendering

One important ability of React is rendering components or data conditionally. We will use this intensively in the next main features that we are going to implement.

Generally, you can accomplish conditional rendering by using the curly brace syntax. An example of an `if` statement is provided here:

```
const [shouldRender, setShouldRender] = useState(false);

return (
  <div className="conditional">
    {(shouldRender === true) && (
      <p>Successful conditional rendering!</p>
    )}
  </div>
)
```

This code is the simplest example of conditional rendering. We have the `shouldRender` variable from the component state, and we use this as our condition. When the condition is `true`, the second part—which is our `Successful conditional rendering!` text—will also render. That is because we are using the `&&` characters. The text does not render if the condition is `false`.

You can replace the preceding condition with anything that you have in mind. It can be a complex condition, such as a function returning a Boolean value, or just as in the preceding code, it can be a state variable.

You will see further examples in later steps and chapters of this book.

Rendering child components

In all of the code that we have written so far, we have directly written the markup as though it is rendered to real **HyperText Markup Language** (**HTML**).

A great feature that React offers is the ability to pass children to other components. The parent component decides what is done with its children.

Something that we are still missing is a good error message for our users. So, we will use this pattern to solve the issue.

Create an `error.js` file next to the `loading.js` file in the `components` folder, as follows:

```
import React from 'react';

export default ({ children }) => {
  return (
    <div className="error message">
      {children}
    </div>
  );
}
```

When passing children to another component, a new property, called `children`, is added to the properties of the component. You specify `children` by writing normal React markup.

If you wanted to, you could perform actions, such as looping through each child. In our example, we render the children as usual, by using curly braces and putting the `children` variable inside.

To start using the new `Error` component, you can simply import it. The markup for the new component is shown here:

```
if (error) return <Error><p>{error.message}</p></Error>;
```

Add some CSS, and everything should be finished, as shown in the following code snippet:

```
.message {
  margin: 20px auto;
  padding: 5px;
  max-width: 400px;
}

.error.message {
  border-radius: 5px;
  background-color: #FFF7F5;
  border: 1px solid #FF9566;
  width: 100%;
}
```

A working result might look like this:

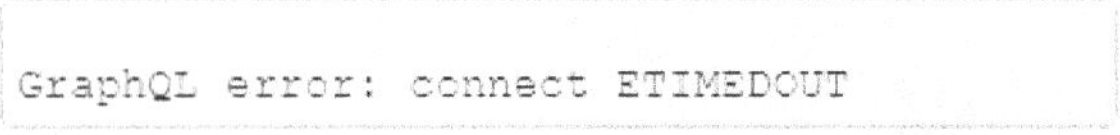

```
GraphQL error: connect ETIMEDOUT
```

Figure 5.1 – Error message

You can apply the stateless function pattern and the children pattern to many other use cases. Which one you use will depend on your specific scenario. In this case, you could also use a stateless function, rather than a React component.

Next, we are going to have a look at how we can improve our code structure, together with the fundamentals that React is following.

Structuring our React application

We have already improved some things by using React patterns. You should do some homework and introduce those patterns wherever possible.

When writing applications, one key objective is to keep them modular and readable but also as understandable as possible. It is always hard to tell when splitting code up is useful and when it overcomplicates things. This is something that you will learn more and more about by writing as many applications and as much code as possible.

Let's begin to structure our application further.

The React file structure

We have already saved our `Loading` and `Error` components in the `components` folder. Still, there are many parts of our components that we did not save in separate files to improve the readability of this book.

I will explain the most important solution for unreadable React code in one example. You can implement this on your own later for all other parts of our application, as you should not read duplicate code.

Currently, we render the posts in our feed by mapping through all posts from the GraphQL response. There, we directly render the corresponding markup for all post items. Therefore, it is one big render function that does everything at once.

To make this a bit more intuitive, we should create a new `Post` component. Separating the components hugely improves the readability of our posts feed. Then, we can replace the return value from the loop with a new component, instead of real markup.

Instead of creating a `post.js` file in our `components` folder, we should first create another `post` folder, as follows:

```
mkdir src/client/components/post
```

The `Post` component consists of multiple tiny, nested components. A post is also a standalone GraphQL entity, making it logical to have a separate folder. We will store all related components in this folder.

Let's create those components. We will start with the post header, where the top part of a post item is defined. Create a new `header.js` file in the `components/post` folder, as follows:

```
import React from 'react';

export default ({post}) => {
  return (
    <div className="header">
      <img src={post.user.avatar} />
      <div>
        <h2>{post.user.username}</h2>
      </div>
    </div>
  );
}
```

The `header` component is just a functional component. As you can see, we are using a React pattern from the earlier pages of this chapter.

Up next is the post content, which represents the body of a post item. Add the following code inside of a new file, called `content.js`:

```
import React from 'react';

export default ({post}) =>
  <p className="content">
    {post.text}
  </p>
```

The code is pretty much the same as that of the post header. At later points, you will be free to introduce real React components or extended markup to those two files. It is entirely open to your implementation.

The main file is a new index.js file in the new post folder. It should look like this:

```
import React from 'react';
import PostHeader from './header';
import PostContent from './content';

const Post = ({ post }) => {
  return (
    <div className={"post " + (post.id < 0 ? "optimistic":
      "") }>
      <PostHeader post={post}/>
      <PostContent post={post}/>
    </div>
  )
}

export default Post
```

The preceding code represents a very basic component, but instead of directly using markup to render a complete post item (as we did before), we are using two further components for this, with PostHeader and PostContent. Both of the components receive post as a property.

You can now use the new Post component in the feed list with ease. First, import the component, as follows:

```
import Post from './components/post';
```

The good thing when having an index.js file is that you can reduce the path by directly pointing to the folder and the index.js file will be picked up automatically. Then, just replace the old code inside the loop, as follows:

```
<Post key={post.id} post={post} />
```

The improvement is that all three of the components give you a clear overview at first glance. Inside of the loop, we return a post item. A post item consists of a header and body content.

Still, there is room for enhancement, because the posts feed list is cluttered.

Efficient Apollo React components

We have successfully replaced the post items in our feed with a React component, instead of raw markup.

A major part, which I dislike very much, is the way we define and pass the actual GraphQL query or mutation to the useQuery or useMutation Hook within the components. It would be good to define those GraphQL requests once and use them everywhere they are needed so that we can use them across different components.

Furthermore, not only should the pure GraphQL requests be reusable but the update function or the optimisticResponse object should also be reusable across different components.

As an example, we will fix those issues for Feed.js in the next section.

Using fragments with Apollo

A GraphQL fragment is a functionality that allows you to share common queries or pieces of them between different bigger GraphQL queries. For example, if you have different GraphQL queries that request a user object and the data structure for this user is always the same, you can define a user fragment to reuse the same attributes across multiple different GraphQL requests.

We will implement this for our GET_POSTS query. Follow the next instructions:

1. Create a new queries folder and a fragments folder inside of the apollo folder, as follows:

    ```
    mkdir src/client/apollo/queries
    mkdir src/client/apollo/fragments
    ```

2. Create a file called userAttributes.js inside the fragments folder and fill in the following lines of code:

    ```
    import { gql } from '@apollo/client';

    export const USER_ATTRIBUTES = gql'
      fragment userAttributes on User {
        username
        avatar
    ```

```
    }
  ';
```

For all `user` objects that we have requested so far, we have always returned the username and the avatar image, as we always display the name in the user profile image. In the preceding code snippet, we implemented the matching GraphQL fragment, which we can now use in any other query where we want to request exactly this data.

3. Create a file called `getPosts.js` inside the `queries` folder. Add the following content to it:

```
import { gql } from '@apollo/client';
import { USER_ATTRIBUTES } from '../fragments/
userAttributes';

export const GET_POSTS = gql'
  query postsFeed($page: Int, $limit: Int) {
    postsFeed(page: $page, limit: $limit) {
      posts {
        id
        text
        user {
          ...userAttributes
        }
      }
    }
  }
  ${USER_ATTRIBUTES}
';
```

In the preceding code snippet, we are making use of our newly created GraphQL fragment. Firstly, we of course import it at the top of the file. Inside the GraphQL query, we use the ...`userAttributes` syntax, which is similar to the normal JavaScript destructuring assignment and spreads the fragment with the same name up to inject those properties at the specified location in the GraphQL query. The last step is to add a fragment below the actual query to be able to make use of it.

4. The last step is to simply replace the old GET_POSTS variable that we have manually parsed inside the Feed.js file with this import statement:

```
import { GET_POSTS } from './apollo/queries/getPosts';
```

At this point, we are successfully using the fragment within our main GET_POSTS query, which we also can reuse just by importing it into any other component. You can repeat this for all the other requests and, by doing so, reach a cleaner code structure and more reusability.

Reusing Apollo Hooks

We did a good job extracting the GraphQL query from our component into separate reusable files.

Still, we have some amount of logic in our component where we define how the update function and optimisticResponse object for the GraphQL mutations are handled. We can also extract these to further improve reusability and clear the code.

One problem with the use of the Apollo or React Hooks is that they need to be executed within the functional component. Still, we can save the biggest amount of the Hook configuration outside of the component to make them reusable.

Follow the next instructions to accomplish this for the addPost mutation:

1. Create a new mutations folder inside the apollo folder, as follows:

```
mkdir src/client/apollo/mutations
```

2. Create a file called addPost.js and insert the following code into it:

```
import { gql } from '@apollo/client';
import { USER_ATTRIBUTES } from '../fragments/
userAttributes';

export const ADD_POST = gql'
  mutation addPost($post : PostInput!) {
    addPost(post : $post) {
      id
      text
      user {
        ...userAttributes
      }
```

```
        }
    }
    ${USER_ATTRIBUTES}
';
```

The great thing is that we can reuse the fragment we have created before here, so we do not need to define it again.

3.   We can use this `ADD_POST` variable in the `Feed.js` file for our mutation. Replace the actual parsed GraphQL query with the following `import` statement:

```
import { ADD_POST } from './apollo/mutations/addPost';
```

4.   We still have the `update` function and the `optimisticResponse` object within our `useMutation` Hook. It's easiest to also move them to the `addPost.js` file, as follows:

```
export const getAddPostConfig = (postContent) => ({
    optimisticResponse: {
        __typename: "mutation",
        addPost: {
            __typename: "Post",
            text: postContent,
            id: -1,
            user: {
                __typename: "User",
                username: "Loading...",
                avatar: "/public/loading.gif"
            }
        }
    },
    update(cache, { data: { addPost } }) {
        cache.modify({
            fields: {
                postsFeed(existingPostsFeed) {
                    const { posts: existingPosts } =
                        existingPostsFeed;
                    const newPostRef = cache.writeFragment({
                        data: addPost,
                        fragment: gql'
```

```
              fragment NewPost on Post {
                  id
                  type
              }
          '
        });
        return {
            ...existingPostsFeed,
            posts: [newPostRef, ...existingPosts]
        };
      }
    }
  });
}
});
```

We are returning a function on import so that we can pass all required parameters. The only expected parameter is the postContent state variable, which is required for the optimisticResponse object.

5. Update the import statement inside the Feed.js file again to also import the getAddPostConfig function, as follows:

```
import { ADD_POST, getAddPostConfig } from './apollo/
mutations/addPost';
```

6. The final useState Hook should then look like this:

```
const [addPost] = useMutation(ADD_POST,
getAddPostConfig(postContent));
```

We execute getAddPostConfig as the second parameter of the useMutation Hook. It will be filled with the returned object but will still have the postContent value inside the optimisticResponse object, as every time the value changes, the getAddPostConfig function will also be run.

7. We can go even further. Add the following line of code to the addPost.js file:

```
export const useAddPostMutation = (postContent) =>
useMutation(ADD_POST, getAddPostConfig(postContent));
```

We will replace the plain `useMutation` Hook in our component and instead build a small wrapper around it. It will accept the `postContent` value to pass it further to the `getAddPostConfig` function. The advantage will be that we can just import the `useAddPostMutation` function and, after executing it, all default configurations are applied and we are able to use that inside any component without having to import the query, the configuration, and the `useMutation` Hook separately.

8. Change the import of the `addPost.js` file inside the `Feed.js` file, as follows:

```
import { useAddPostMutation } from './apollo/mutations/
addPost';
```

9. Replace the `useMutation` Hook with the following line of code and remove the `useMutation` import while doing so:

```
const [addPost] = useAddPostMutation(postContent);
```

We only need to import one function to get the full-fledged `addPost` mutation running in any component we want it to be in. This concept will work with any query or mutations we have used so far.

We cleaned up the functional component from most of the GraphQL request logic and have it in separate files now. We can use the query or the mutation at any location where we need them.

I recommend that you repeat the same for all other locations so that you have a nicely filled set of GraphQL requests that we are able to reuse anywhere we want. This would be a good homework exercise to learn that concept.

Next, we will have a look at how we can extend Graphbook.

Extending Graphbook

Our social network is still a bit rough. Aside from the fact that we are still missing authentication, all of the features are pretty basic; writing and reading the posts and messages is nothing exceptional.

If you compare it to Facebook, there are many things that we need to do. Of course, we cannot rebuild Facebook in its totality, but the usual features should be there. From my point of view, we should cover the following features:

- Adding a drop-down menu to the posts to allow deletion of posts.
- Creating a global `user` object with the React Context API.

- Using Apollo cache as an alternative to the React Context API.

- Implementing a top bar as the first component rendered above all of the views. We can search for users in our database from a search bar, and we can show the logged-in user from the global user object.

We will begin by looking at the first feature.

The React context menu

You should be able to write the React context menu pretty much on your own. All the required React patterns have been explained, and implementing the mutations should be clear by now.

Before we begin, we will lay out the plan that we want to follow. We'll aim to do this:

- Render a simple icon with Font Awesome

- Build React helper components

- Handle the onClick event and set the correct component state

- Use the conditional rendering pattern to show the drop-down menu, if the component state is set correctly

- Add buttons to the menu and bind mutations to them

Continue reading to find out how to get the job done.

The following is a preview screenshot, showing how the final implemented feature should look:

Figure 5.2 – Drop-down with context

We will now start with the first task of setting up Font Awesome for our project.

Font Awesome in React

As you may have noticed, we have not installed Font Awesome yet. Let's fix this with npm, as follows:

```
npm i --save @fortawesome/fontawesome-svg-core @fortawesome/
free-solid-svg-icons @fortawesome/free-brands-svg-icons @
fortawesome/react-fontawesome
```

Graphbook relies on the preceding four packages to import the Font Awesome icons into our frontend code.

> **Important Note**
>
> Font Awesome provides multiple configurations for use with React. The best, most production-ready approach is to import only the icons that we are explicitly going to use. For your next project or prototype, it might make sense to get started with the simplest approach. You can find all of the information on the official page, at https://fontawesome.com/v5.15/how-to-use/on-the-web/using-with/react.

Creating a separate file for Font Awesome will help us to have a clean import. Save the following code under the fontawesome.js file, inside of the components folder:

```
import { library } from '@fortawesome/fontawesome-svg-core';
import { faAngleDown } from '@fortawesome/free-solid-svg-
icons';


library.add(faAngleDown);
```

First, we import the library object from the Font Awesome core package. For our specific use case, we only need one arrow image, called angle-down. Using the library.add function, we register this icon for later use.

> **Important Note**
>
> There are many versions of Font Awesome. In this book, we are using Font Awesome 5, with free icons only. More premium icons can be bought on the official Font Awesome web page. You can find an overview of all of the icons, and a detailed description of each, in the icon gallery at https://fontawesome.com/icons?d=gallery.

The only place where we need this file is within our root App.js file. It ensures that all of our custom React components can display the imported icons. Add the following import statement to the top of the file:

```
import './components/fontawesome';
```

No variable is required to save the exported methods since there won't be any. We want to execute this file in our application only once.

When you reach a point where your application needs a complete set of icons, you can get all the icons grouped directly from the @fortawesome/free-brands-svg-icons package, which we also installed.

Next, we are going to create a Dropdown helper component.

React helper components

Production-ready applications need to be as polished as possible. Implementing reusable React components is one of the most important things to do.

You should notice that drop-down menus are a common topic when building client-side applications. They are global parts of the frontend and appear everywhere throughout our components.

It would be best to separate the actual menu markup that we want to display from the code, which handles the event-binding and shows the menu.

I always call this kind of code in React **helper components**. They are not implementing any business logic but give us the opportunity to reuse drop-down menus or other features wherever we want.

Logically, the first step is to create a new folder to store all of the helper components, as follows:

```
mkdir src/client/components/helpers
```

Create a new file, called dropdown.js, as the helper component, as follows:

```
import React, { useState, useRef, useEffect } from 'react';

export default ({ trigger, children }) => {
  const [show, setShow] = useState(false);
  const wrapperRef = useRef(null);
```

```
useOutsideClick(wrapperRef);

function useOutsideClick(ref) {
  useEffect(() => {
    function handleClickOutside(event) {
      if (ref.current &&
          !ref.current.contains(event.target)) {
        setShow(false);
      }
    }

    document.addEventListener("mousedown",
      handleClickOutside);
    return () => {
      document.removeEventListener("mousedown",
        handleClickOutside);
    };
  }, [ref]);
}

return(
  <div className="dropdown">
    <div>
      <div className="trigger" onClick={() =>
        setShow(!show)}>
        {trigger}
      </div>
      <div ref={wrapperRef}>
        { show &&
          <div className="content">
            {children}
          </div>
        }
      </div>
    </div>
```

```
        </div>
    )
}
```

We do not require much code to write a drop-down component. It is also pretty efficient since this works with nearly every scenario that you can think of.

We use basic event handling in the preceding code snippet. When the trigger `div` tag is clicked, we update the `show state` variable. Inside of the `div` trigger, we also render a property called `trigger`. A `trigger` property can be anything from regular text or an HTML tag to a React component. It can be passed through the parent components in order to customize the look of the drop-down component.

In addition to the `trigger` property, we are using two well-known React patterns, as follows:

- Conditional rendering when the `show` variable is `true`
- Rendering children given by the parent component

This solution allows us to fill in the menu items that we want to render directly as children of the `Dropdown` component, which, as mentioned previously, is displayed after clicking on the trigger. In this case, the `show` state variable is `true`.

However, one thing is still not completely correct here. If you test the drop-down component by providing simple text or an icon as a trigger and other text as the content, you should see that the `Dropdown` component only closes when clicking on the trigger again; it does not close when clicking anywhere else in our browser, outside of the drop-down menu.

This is one scenario where the React approach encounters problems. There is no DOM Node event such as `useOutsideClick`, so we cannot directly listen to the outside click events of any DOM Node, such as our drop-down menu. The conventional approach is to bind an event listener to the complete document. Clicking anywhere in our browser closes the drop-down menu.

The `useOutSideClick` Hook just checks if the clicked element matches the reference that we set up through the `useRef` Hook.

When clicking on the trigger button, we add the click event listener to the whole document with the `addEventListener` function of JavaScript.

> **Important Note**
>
> There are many cases where it might make sense to leave the React approach and use the DOM directly, through the standard JavaScript interface.
>
> Read this article on *Medium* to get a better understanding: `https://` `medium.com/@garrettmac/reactjs-how-to-safely-` `manipulate-the-dom-when-reactjs-cant-the-right-` `way-8a20928e8a6`.

The `useEffect` Hook is executed only on first rendering of the component. You can execute any kind of logic there. If the return value is a function too, this function will be executed on unmount of the component. By doing that, we do not forget to remove all the manually created event listeners whenever the component is unmounted and removed from the DOM. Forgetting this can lead to many errors.

As mentioned previously, this is the part where React fails at least a little bit, although it is not the fault of React. The DOM and JavaScript do not have the right abilities.

We can finally use our helper component and display the context menus for posts, but first, we need to prepare all of the menu items and components that we want to render.

The Apollo deletePost mutation

A mutation is always located at two points in our code. One part is written inside of our GraphQL API in the backend, and the other one is written in our frontend code.

We should start with the implementation on the backend side, as follows:

1. Edit the GraphQL schema. The `deletePost` mutation needs to go inside of the `RootMutation` object. The new `Response` type serves as a return value, as deleted posts cannot be returned because they do not exist. Note in the following code snippet that we only need the `postId` parameter and do not send the complete post:

```
type Response {
  success: Boolean
}

deletePost (
  postId: Int!
): Response
```

2. Add the missing GraphQL resolver function. The `destroy` function of Sequelize only returns a number that represents the number of deleted rows. We return an object with the `success` field. This field indicates whether our frontend should throw an error. The code is illustrated in the following snippet:

```
deletePost(root, { postId }, context) {
  return Post.destroy({
    where: {
      id: postId
    }
  }).then(function(rows) {
    if(rows === 1) {
      logger.log({
        level: 'info',
        message: 'Post ' + postId + 'was deleted',
      });
      return {
        success: true
      };
    }
    return {
      success: false
    };
  }, function(err) {
    logger.log({
      level: 'error',
      message: err.message,
    });
  });
},
```

The only special thing here is that we need to specify which posts we want to delete. This is done by having the `where` property inside of the function call. Because we currently do not have authentication implemented yet, we cannot verify the user deleting the post, but for our example, this is no problem.

In short, our GraphQL API is now able to accept the `deletePost` mutation. We do not verify which user sends this mutation, so for our example, posts can be deleted by anyone.

We can now focus on the frontend again.

Recall how we implemented the previous mutations; we always created reusable functions and configurations for them. We should do the same for the delete mutation.

Let's start by implementing the deletePost mutation for the client, as follows:

1. Create a new file, called deletePost.js, inside of the mutations folder.

2. Import all the dependencies, as follows:

```
import { gql } from '@apollo/client';
import { useMutation } from '@apollo/client';
```

3. Add the new deletePost mutation, as follows:

```
export const DELETE_POST = gql'
  mutation deletePost($postId : Int!) {
    deletePost(postId : $postId) {
      success
    }
  }
';
```

4. Add a new function to handle the configuration of the mutation, as follows:

```
export const getDeletePostConfig = (postId) => ({
  update(cache, { data: { deletePost: { success } } })
  {
    if(success) {
      cache.modify({
        fields: {
          postsFeed(postsFeed, { readField }) {
            return {
              ...postsFeed,
              posts: postsFeed.posts.filter(postRef =>
              postId !== readField('id', postRef))
            }
          }
        }
      })
    }
  }
});
```

```
        }
      }
   });
```

To remove an item, we need to clean the array from the post with the given `postId` value. The easiest way to do that is to return the complete object of `postsFeed`, and while doing so, we let the normal JavaScript `filter` function return only those posts not having the `postId` value inside. To read the **identifier** (**ID**) of the current post, we use the `readField` function given by Apollo.

5. Lastly, insert the wrapper function for the `useMutation` Hook, as follows:

```
export const useDeletePostMutation = (postId) =>
useMutation(DELETE_POST, getDeletePostConfig(postId));
```

I have removed the `optimisticResponse` update for the **user interface** (**UI**) since it is not intuitive if the request fails because then the UI would first show an optimistic update, but under failure, put the post back again as the API call failed. This would make your post disappear and reappear again.

We need to add our drop-down menu to the post header with a new item so that we can call the `deletePost` mutation. Follow these instructions to add it:

1. Open the `header.js` file and import the drop-down component, `fontawesome`, and the mutation, as follows:

```
import Dropdown from '../helpers/dropdown';
import { FontAwesomeIcon } from '@fortawesome/react-
fontawesome';
import { useDeletePostMutation } from '../../apollo/
mutations/deletePost';
```

2. Run the `useDeletePostMutation` Hook before the `return` statement, as follows:

```
const [deletePost] = useDeletePostMutation(post.id);
```

3. Add the new button to our header beneath the `div` tag with the username, like this:

```
<Dropdown trigger={<FontAwesomeIcon icon="angle-down"
/>}>
  <button onClick={() => deletePost({ variables: {
    postId: post.id }})}>Delete</button>
</Dropdown>
```

The overall solution is very simple. We have a wrapping drop-down component. All children will be only rendered if the show state variable is changed. This includes our **Delete** button. The **Delete** button just receives the deletePost mutation and triggers on click. The mutation itself is separated from the actual code to render the view.

You can get the correct CSS from the GitHub repository.

We have now covered the retrieval and deletion of posts. The update of posts is more or less the same—instead of adding or removing, you need to update the post not only via its ID in the database but also in the Apollo cache. The approach is the same, though.

I expect that you are now prepared for advanced scenarios, where communication between multiple components on different layers is required. Consequently, when starting the server and client, you should be presented with the preview image that I gave you when starting this section.

To get some more practice, we will repeat this for another use case in the next section.

The React application bar

In contrast with Facebook, we do not have an outstanding application bar. The plan is to implement something similar. It is fixed to always stay at the top of the browser window, above all parts of the Graphbook. You will be able to search for other users, see notifications, and see the logged-in user inside the application bar, after going through this section.

The first thing that we will implement is a simple search for users because it is complex.

The following screenshot shows a preview of what we are going to build:

Figure 5.3 – Search results

It looks basic, but what we are doing here is binding the onChange event of an input and refetching the query every time the value changes. Logically, this rerenders the search list in accordance with the responses from our GraphQL API.

Starting with the API, we need to introduce a new entity.

Just as with our `postsFeed` query, we will set up pagination from the beginning, because later, we might want to offer more advanced functionalities, such as loading more items while scrolling through the search list.

Edit the GraphQL schema and fill in the new `RootQuery` property and type, as follows:

```
type UsersSearch {
  users: [User]
}
```

```
usersSearch(page: Int, limit: Int, text: String!): UsersSearch
```

The `UsersSearch` type expects one special parameter, which is the search text. Without the text parameter, the request would not make much sense. You should remember the `page` and `limit` parameters from the `postsFeed` pagination.

Furthermore, the `resolver` function looks pretty much the same as the `postsFeed` resolver function. You can add the following code straight into the `resolvers.js` file in the `RootQuery` property, as follows:

```
usersSearch(root, { page, limit, text }, context) {
  if(text.length < 3) {
    return {
      users: []
    };
  }
  var skip = 0;
  if(page && limit) {
    skip = page * limit;
  }
  var query = {
    order: [['createdAt', 'DESC']],
    offset: skip,
  };
  if(limit) {
    query.limit = limit;
  }
  query.where = {
    username: {
```

```
        [Op.like]: '%' + text + '%'
      }
    };
    return {
      users: User.findAll(query)
    };
  },
```

You should note that the first condition asks whether the provided text is larger than three characters. We do this to avoid sending too many unnecessary queries to our database. Searching for every user where the username consists of just one or two characters would result in providing us with nearly every user. Of course, this could have been done on the frontend, too, but various clients could use our API, so we need to make sure that the backend makes this small improvement as well.

We send the `query` object to our database through Sequelize. The code works pretty much like the `postsFeed` resolver function from before, except that we are using a Sequelize operator. We want to find every user where the username includes the entered text, without specifying whether it is at the start, middle, or end of the name. Consequently, we will use the `Op.like` operator, which Sequelize parses into a pure **Structured Query Language** (**SQL**) `LIKE` query, giving us the results we want. The `%` operator is used in MySQL to represent an unspecified number of characters. To enable this operator, we must import the `sequelize` package and extract the `Op` object from it, as follows:

```
import Sequelize from 'sequelize';
const Op = Sequelize.Op;
```

Going further, we can implement the client-side code, as follows:

1. Create a file called `searchQuery.js` within the `queries` folder and insert the following code into it:

```
import { gql } from '@apollo/client';
import { USER_ATTRIBUTES } from '../fragments/
userAttributes';
import { useQuery } from '@apollo/client';

export const GET_USERS = gql'
  query usersSearch($page: Int, $limit: Int, $text:
    String!) {
```

```
    usersSearch(page: $page, limit: $limit, text:
      $text) {
      users {
        id
        ...userAttributes
      }
    }
  }
  ${USER_ATTRIBUTES}
';


export const getUserSearchConfig = (text) => ({
variables: { page: 0, limit: 5, text }, skip: text.length
< 3})


export const useUserSearchQuery = (text) => useQuery(GET_
USERS, getUserSearchConfig(text))
```

The only required parameter is the `text` parameter that we want to pass to our search. If the `text` parameter is shorter than three characters, we are passing a `skip` property either with `true` or `false` to not execute the GraphQL request.

2. Continuing with our plan, we will create an application bar in a separate file. Create a new folder, called `bar`, below the `components` folder and the `index.js` file. Fill it in with the following code:

```
import React from 'react';
import SearchBar from './search';


const Bar = () => {
  return (
    <div className="topbar">
      <div className="inner">
        <SearchBar/>
      </div>
    </div>
  );
```

```
    }

    export default Bar
```

This file works as a wrapper for all of the components we want to render in the application bar; it does not implement any custom logic. We have already imported the `SearchBar` component that we must create.

3. The `SearchBar` component lives inside of a separate file. Just create a `search.js` file in the `bar` folder, as follows:

```
import React, { useState } from 'react';
import { useUserSearchQuery } from '../../apollo/queries/
searchQuery';
import SearchList from './searchList';

const SearchBar = () => {
  const [text, setText] = useState('');
  const { loading, error, data } =
    useUserSearchQuery(text);

  const changeText = (event) => {
    setText(event.target.value);
  }

  return (
    <div className="search">
      <input type="text" onChange={changeText}
        value={text}
      />
      {!loading && !error && data && (
        <SearchList data={data}/>
      )}
    </div>
  );
}

export default SearchBar
```

We are storing the current input value inside of a state variable, called `text`. Every time the text is changed, the `useUserSearchQuery` Hook is executed again with the new `text` parameter. Inside of the query Hook, the value is merged into the variables and sent with a GraphQL request. The result is then handed over to the `SearchList` component inside the `data` property if the request is not loading and did not have any error.

4. Next, we will implement the `SearchList` component. This behaves like the posts feed, but only renders something if a response is given with at least one user. The list is displayed as a drop-down menu and is hidden whenever the browser window is clicked on. Create a file called `searchList.js` inside of the `bar` folder, with the following code:

```
import React, { useState, useEffect } from 'react';

const SearchList = ({ data: { usersSearch: { users }}})
=> {
  const [show, setShowList] = useState(false);

  const handleShow = (show) => {
    if(show) {
      document.addEventListener('click',
        handleShow.bind(null, !show), true);
    } else {
      document.removeEventListener('click',
        handleShow.bind(null, !show), true);
    }
    setShowList(show);
  }

  const showList = (users) => {
    if(users.length) {
      handleShow(true);
    } else {
      handleShow(false);
    }
  }
}
```

```
useEffect(() => {
  showList(users);
}, [users]);

useEffect(() => {
  return () => {
    document.removeEventListener('click',
      handleShow.bind(null, !show), true);
  }
});

return (
  show &&
    <div className="result">
      {users.map((user, i) =>
        <div key={user.id} className="user">
          <img src={user.avatar} />
          <span>{user.username}</span>
        </div>
      )}
    </div>
  )
}

export default SearchList
```

We are using the useEffect Hook here with the dependency on users, which is executed whenever the parent component sets new properties on the current one. In this case, we check whether the properties include at least one user, and then set the state accordingly, in order to make the drop-down menu visible. The drop-down menu is hidden when clicked on or when an empty result is given. The approach is very similar to the drop-down for the post.

There are just two things to do now, as follows:

1. You should copy the CSS from the official GitHub repository of this chapter in order to get the correct styling, or you can do it on your own.

2. You need to import the bar wrapper component inside of the App class and render it between React Helmet and the news feed.

The first feature of our application bar is now complete.

Let's continue and take a look at React's Context API, the Apollo Consumer feature, and how to store data globally in our React frontend.

The React Context API versus Apollo Consumer

There are two ways to handle global variables in the stack that we are using at the moment. These are through the new React Context API and the Apollo Consumer functionality.

> **Important Note**
>
> There are further ways to handle global state management. One of the most famous libraries is Redux, but there are many more. As an explanation of Redux would outreach the topic of this book, we are only focusing on the tools provided by React and Apollo.
>
> If you want to check other ways, have a look at Redux's website: `https://redux.js.org/`.

From version 16.3 of React, there is a Context API that allows you to define global providers offering data through deeply nested components. These components do not require your application to hand over the data through many components, from the top to the bottom of the React tree. Instead, it uses so-called consumers and providers. These are useful when you set up the `user` object at a global point of your application, and you can access it from anywhere. In earlier versions of React, you needed to pass the property down from component to component to get it to the correct component at the bottom of the React component tree. This passing of properties through multiple layers of components is also called "prop drilling."

An alternative approach to the React Context API is the Apollo Consumer feature, which is a specific implementation for Apollo. The React Context API is a general way of doing things, for Apollo or anything else that you can imagine.

The great thing about the Apollo Consumer component is that it enables you to access the Apollo cache and use it as data storage. Using the Apollo Consumer component saves you from handling all the data, and you are also not required to implement the provider itself; you can consume the data wherever you want.

Both approaches will result in the following output:

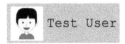

Figure 5.4 – User profile in the top bar

The best option is to show you the two alternatives right away so that you can identify your preferred method.

The React Context API

We will start with the React method for storing and accessing global data in your frontend.

Here is a short explanation of this method:

- **Context**: This is a React approach for sharing data between components, without having to pass it through the complete tree.

- **Provider**: This is a global component, mostly used at just one point in your code. It enables you to access the specific context data.

- **Consumer**: This is a component that can be used at many different points in your application, reading the data behind the context that you are referring to.

To get started, create a folder called `context` below the `components` folder. In that folder, create a `user.js` file, where we can set up the Context API.

We will go over every step, one by one, as follows:

1. As always, we need to import all of the dependencies. Furthermore, we will set up a new empty context. The `createContext` function will return one provider and consumer to use throughout the application, as follows:

   ```
   import React, { createContext } from 'react';
   const { Provider, Consumer } = createContext();
   ```

2. Now, we want to use the provider. The best option here is to create a special `UserProvider` component. Later, when we have authentication, we can adjust it to do the GraphQL query, and then share the resultant data in our frontend. For now, we will stick with fake data. Insert the following code:

   ```
   export const UserProvider = ({ children }) => {
     const user = {
       username: "Test User",
   ```

```
        avatar: "/uploads/avatar1.png"
    };
    return (
        <Provider value={user}>
            {children}
        </Provider>
    );
}
```

3. In the preceding code snippet, we render the `Provider` component from Apollo and wrap all the children in it. There is a `Consumer` component that reads from the `Provider`. We will set up a special `UserConsumer` component that takes care of passing the data to the underlying components by cloning them with React's `cloneElement` function, as follows:

```
export const UserConsumer = ({ children }) => {
    return (
        <Consumer>
            {user => React.Children.map(children,
                function(child){
                return React.cloneElement(child, { user });
            })}
        </Consumer>
    )
}
```

We will export both classes directly under their names.

We need to introduce the provider at an early point in our code base. The best approach is to import the `UserProvider` component into the `App.js` file, as follows:

```
import { UserProvider } from './components/context/user';
```

Use the provider as follows, and wrap it around all essential components:

```
<UserProvider>
    <Bar />
    <Feed />
    <Chats />
</UserProvider>
```

Everywhere in the Bar, Feed, and Chats components, we can now read from the provider.

As stated previously, we want to show the logged-in user, with their name, inside the application.

The component using the data is the UserBar component. We need to create a user.js file inside of the bar folder. Insert the following code:

```
import React from 'react';

const UserBar = ({ user }) => {
  if(!user) return null;

  return (
    <div className="user">
      <img src={user.avatar} />
      <span>{user.username}</span>
    </div>
  );
}

export default UserBar
```

For the moment, we render a simple user container inside of the application bar, from the data of the user object.

To get the user data into the UserBar component, we need to use the UserConsumer component, of course.

Open the index.js file for the top bar and add the following code to the return statement, beneath the SearchBar component:

```
<UserConsumer>
  <UserBar />
</UserConsumer>
```

Obviously, you need to import both of the components at the top of the file, as follows:

```
import UserBar from './user';
import { UserConsumer } from '../context/user';
```

You have now successfully configured and used the React Context API to save and read data globally.

The solution that we have is a general approach that will work for all scenarios that you can think of, including Apollo. If you view the browser now, you will be able to see the logged-in user or at least the fake data we added in the top bar.

Nevertheless, we should cover the solution offered by Apollo itself.

Apollo Consumer

Nearly all of the code that we have written can stay as it was in the previous section. We just need to remove the `UserProvider` component from the `App` class because it is not needed anymore for the Apollo Consumer component.

Open up the `user.js` file in the `context` folder and replace the contents with the following code:

```
import React from 'react';
import { ApolloConsumer } from '@apollo/client';

export const UserConsumer = ({ children }) => {
  return (
    <ApolloConsumer>
      {client => {
        // Use client.readQuery to get the current logged
        // in user.
        const user = {
          username: "Test User",
          avatar: "/uploads/avatar1.png"
        };
        return React.Children.map(children,
          function(child){
          return React.cloneElement(child, { user });
```

```
          });
      }}
   </ApolloConsumer>
 )
}
```

As you can see, we import the `ApolloConsumer` component from the `@apollo-client` package. This package enables us to get access to the Apollo Client that we set up in *Chapter 4*, *Hooking Apollo into React*.

The problem we have here is that we do not have a `CurrentUser` query that would respond with the logged-in user from the GraphQL, so we are not able to run the `readQuery` function. You would typically run the query against the internal cache of Apollo and be able to get the `user` object easily. Once we have implemented authentication, we will fix this problem.

For now, we will return the same fake object as we did with the React Context API. The Apollo Client replaces the `Provider` component that we used with the React Context API.

I hope that you can understand the difference between these two solutions. In the next chapter, you will see the `ApolloConsumer` component in full action, when the user query is established and can be read through the client of its cache.

Documenting React applications

We have put a lot of work and code into our React application. To be honest, we can improve upon our code base by documenting it. We did not comment on our code, we did not add React component property-type definitions, and we have no automated documentation tool. Of course, we did not write any comments because you learned all of the techniques and libraries from the book, so no comments were needed. However, be sure to always comment your code outside of this book.

In the JavaScript ecosystem, many different approaches and tools exist to document your application. For this book, we will use a tool called **React Styleguidist**. It was made especially for React. You cannot document other frameworks or code with it.

> **Important Note**
>
> Generally speaking, this is an area that you can put months of work into without coming to a real end. If you are searching for a general approach for any framework or backend and frontend, I can recommend JSDoc, but there are many more.
>
> Beyond that, there are many different React documentation tools. If you want to check other tools, have a look here: `https://blog.bitsrc.io/6-tools-for-documenting-your-react-components-like-a-pro-5027cdfb40c6`.

Let's get started with the configuration for React Styleguidist.

Setting up React Styleguidist

React Styleguidist and our application rely on `webpack`. Just follow these instructions to get a working copy of it:

1. Install React Styleguidist using npm, as follows:

    ```
    npm install --save-dev react-styleguidist
    ```

2. Usually, the folder structure is expected to be `src/components`, but we have a `client` folder between the `src` and `components` folders, so we must configure React Styleguidist to let it understand our folder structure. Create a `styleguide.config.js` file in the root folder of the project to configure it, as follows:

```js
const path = require('path')
module.exports = {
  components: 'src/client/components/**/*.js',
  require: [
    path.join(__dirname, 'assets/css/style.css')
  ],
  webpackConfig: require('./webpack.client.config')
}
```

We export an object containing all the information needed for React Styleguidist. In addition to specifying the `components` path, we also require our main CSS style file. You will see why this can be useful later in this chapter. We must define the `webpackConfig` option because our `config` file has a custom name that is not found automatically.

Styleguidist provides two ways to view the documentation. One is to build the documentation statically, in production mode, with this command:

```
npx styleguidist build
```

This command creates a `styleguide` folder, and inside the folder there are HTML files for our documentation. It is an excellent method when releasing new versions of your application so that you can save and back up those files with each version.

> **Note**
>
> If you see an error while running `npx styleguidist`, you have to apply a short-term workaround.
>
> Install yarn by running `npm install -g yarn` and then add the following lines to the root level of your `package.json` object:
>
> ```
> "resolutions": {
>     "react-dev-utils": "12.0.0-next.47"
> }
> ```
>
> You can then run `yarn install`. This will update an internal dependency of `styleguidist` to a newer version that does not have the problem.
>
> You can then run `npx styleguidist build` again. Just remember that in case you run `npm install`, it will overwrite this dependency with the old one and you will have to run `yarn install` again to get it working.

The second method, for development cases, lets Styleguidist run and create documentation on the fly using `webpack`. Here's the command you need for this:

```
npx styleguidist server
```

You can view the results under `http://localhost:6060`. The documentation should look like this:

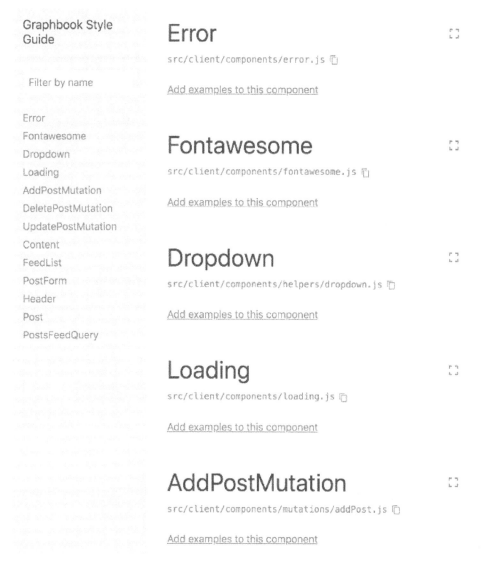

Figure 5.5 – React Styleguidist documentation

In the left-hand panel, all of the components are listed in the order of our folder structure. You will always have an excellent overview of the existing components this way.

In the main panel, each component is explained in detail. You may have noticed that the components are missing further information. We will change that next.

React PropTypes

An essential feature of React is passing the properties to the child components. These can be anything from basic strings to numbers, but also complete components. We have already seen all of the scenarios in our application.

Developers that are new to your code base need to read through all of the components and identify which properties they can accept.

React offers a way to describe properties from within each component. Documenting the properties of your components makes it easier for other developers to understand your React components.

We will take a look at how to do this with an example in our `Post` component.

There are two React features that we haven't covered yet, as follows:

- If your components have optional parameters, it can make sense to have default properties in the first place. To do this, you can specify `defaultProps` as a static property, in the same way as with the state initializers.

- The important part is the `propTypes` field, which you can fill in for all of your components with the custom properties that they accept.

A new package is required to define the property types, as follows:

```
npm install --save prop-types
```

This package includes everything that we need to set up our property definitions.

Now, open your `Post` component's `index.js` file. We need to import the new package at the top of this file, as follows:

```
import PropTypes from 'prop-types';
```

Next, we will add the new field to our component, before the `export` statement, like this:

```
Post.propTypes = {
  /** Object containing the complete post. */
  post: PropTypes.object.isRequired,
}
```

The preceding code should help everyone to understand your component a bit better. Every developer should know that a `post` object is required for this component to work.

The `PropTypes` package offers various types that we can use. You can access each type with `PropTypes.X`. If it is a required property, you can append `isRequired` in the same way as in the preceding code.

Not only does React now throw an error inside of our console when the property does not exist, but React Styleguidist is also able to show which properties are needed, as you can see in the following screenshot:

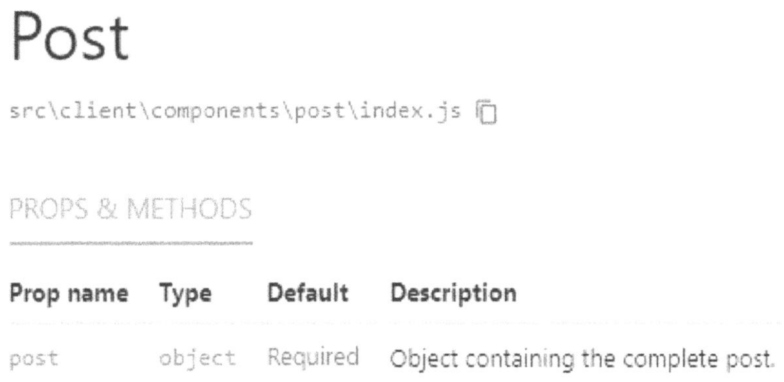

Post

`src\client\components\post\index.js`

PROPS & METHODS

Prop name	Type	Default	Description
post	object	Required	Object containing the complete post.

Figure 5.6 – Basic property documentation

However, what is a `post` object? What kind of fields does it include?

The best way to document a `post` object is to define which properties a post should include, at least for this specific component. Replace the property definition, as follows:

```
Post.propTypes = {
  /** Object containing the complete post. */
  post: PropTypes.shape({
    id: PropTypes.number.isRequired,
    text: PropTypes.string.isRequired,
    user: PropTypes.shape({
      avatar: PropTypes.string.isRequired,
      username: PropTypes.string.isRequired,
    }).isRequired
  }).isRequired,
}
```

Here, we use the shape function. This allows you to hand over a list of fields that the object contains. Each of those is given a type from the PropTypes package.

The output from React Styleguidist now looks like this:

Figure 5.7 – Detailed property documentation

All the fields that we specified are listed separately. At the time of writing this book, React Styleguidist does not offer a recursive view of all properties. As you can see, the user object inside of the post object is not listed with its properties, but it is only listed as a second shape. If you need this feature, you can, of course, implement it yourself and send a pull request on the official GitHub repository, or switch to another tool.

> **Important Note**
> React offers way more property types and functions that you can use to document all of the components and their properties. To learn a bit more about this, visit the official documentation at https://reactjs.org/docs/typechecking-with-proptypes.html.

One last great feature of React Styleguidist is that you can enter examples for every component. You can also use markdown to add some more descriptions.

For our `Post` component, we need to create an `index.md` file, next to the `index.js` file in the `post` folder. React Styleguidist proposes creating either a `Readme.md` or a `Post.md` file, but those did not work for me. The `index.md` file should look like this:

```
Post example:

'''js
  const post = {
    id: 3,
    text: "This is a test post!",
    user: {
      avatar: "/uploads/avatar1.png",
      username: "Test User"
    }
  };


  <Post key={post.id} post={post} />
'''
```

Sadly, you will not be able to see the output of that markup directly. The reason is that the `Post` component relies on Apollo. If you just render the plain `Post` component in the way React Styleguidist does, the Apollo Client will not be there.

To fix this issue, we can overwrite the default way React Styleguidist renders any component. Follow these instructions to get it working:

1. Create a new folder where we can save all special React Styleguidist components, as follows:

   ```
   mkdir src/client/styleguide/
   ```

2. Create a file called `Wrapper.js` with the following content:

   ```
   import React from 'react';
   import client from '../apollo';
   import { ApolloProvider } from '@apollo/client/react';

   const Wrapper = ({ children }) => {
     return (
       <ApolloProvider client={client}>
   ```

```
            {children}
        </ApolloProvider>
    );
}


export default Wrapper
```

This will be the standard `Wrapper` component for all components that React Styleguidist runs. This way, we ensure that we always have the Apollo Client in the context.

3. The last thing we need to do is to add the following property to the `styleguide.config.js` file:

```
styleguideComponents: {
    Wrapper: path.join(__dirname,
      'src/client/styleguide/Wrapper')
},
```

React Styleguidist will now use this `Wrapper` component.

If you restart React Styleguidist, it will render the documentation and generate the following output:

Figure 5.8 – React Styleguidist example

Now, you can see why it was useful to use the CSS style. Not only can React Styleguidist document the code, but it can also execute it within the documentation. As in the preceding code, providing the correct properties inside of the `post` object enables us to see how the component should look, including the correct styling.

This example shows how reusable our `Post` component is since it is usable without having to run the Apollo query.

The basics should be clear by now. Continue to read up on this topic because there are more things to learn.

Summary

Through this chapter, you have gained a lot of experience in writing a React application. You have applied multiple React patterns to different use cases, such as children passing through a pattern and conditional rendering. Furthermore, you now know how to document your code correctly.

You also learned how to use the React Context API, in comparison with the Apollo Consumer feature, to retrieve the currently logged-in user in our application.

In the next chapter, you will learn how to implement authentication in your backend and use it in the frontend.

6
Authentication with Apollo and React

We have come a long way over the last few chapters. We have now reached the point where we are going to implement authentication for our React and GraphQL web applications. In this chapter, you are going to learn about some essential concepts for building an application with authentication using GraphQL.

This chapter covers the following topics:

- What is a JWT?
- Cookies versus localStorage
- Implementing authentication in Node.js and Apollo
- Signing up and logging in users
- Authenticating GraphQL queries and mutations
- Accessing the user from the request context

Technical requirements

The source code for this chapter is available in the following GitHub repository:

```
https://github.com/PacktPublishing/Full-Stack-Web-Development-
with-GraphQL-and-React-Second-Edition/tree/main/Chapter06
```

What are JSON Web Tokens?

JSON Web Tokens (**JWTs**) are still a pretty new standard for carrying out authentication; not everyone knows about them, and even fewer people use them. This section does not provide a theoretical excursion through the mathematical or cryptographic basics of JWTs.

In traditional web applications written using PHP, for example, you commonly have a session cookie. This cookie identifies the user session on the server. The session must be stored on the server to retrieve the initial user. The problem here is that the overhead of saving and querying all the sessions for all the users can be high. When using JWTs, however, there is no need for the server to preserve any kind of session ID.

Generally speaking, a JWT consists of everything you need to identify a user. The most common approach is to store the creation time of the token, the username, the user ID, and maybe the role, such as an admin or a normal user. You should not include any personal or critical data for security reasons.

The reason a JWT exists is not to encrypt or secure data in any way. Instead, to authorize yourself using a resource such as a server, you send a signed JWT that your server can verify. It can only verify the JWT if it was created by a service stated as authentic by your server. In most cases, your server will have used its public key to sign the token. Any person or service that can read the communication between you and the server can access the token and can extract the payload without further ado. They are not able to edit its content though, because the token is signed with a signature.

The token needs to be transported and stored securely in the browser of the client. If the token gets into the wrong hands, that person can access the affected application with your identity, initiate actions in your name, or read personal data. It is also hard to invalidate a JWT. With a session cookie, you can delete the session on the server, and the user will no longer be authenticated through the cookie. With a JWT, however, we do not have any information on the server. It can only validate the signature of the token and find the user in your database. One common approach is to have a blacklist of all the disallowed tokens. Alternatively, you can keep the lifetime of a JWT low by specifying the expiration date. This solution, however, requires the user to frequently repeat the login process, which makes the experience less comfortable.

JWTs do not require any server-side storage. The great thing about server-side sessions is that you can store specific application states for your user and, for example, remember the last actions a user performed. Without a server-side store, you either need to implement these features in `localStorage` or implement a session store, which is not required for using JWT authentication at all:

Note

JWTs are an important topic in developer communities. There is some excellent documentation available related to what JWTs are, how they can be used, and their technological background. Visit the following web page to learn more and to see a demonstration of the generation of a JWT: `https://jwt.io/`.

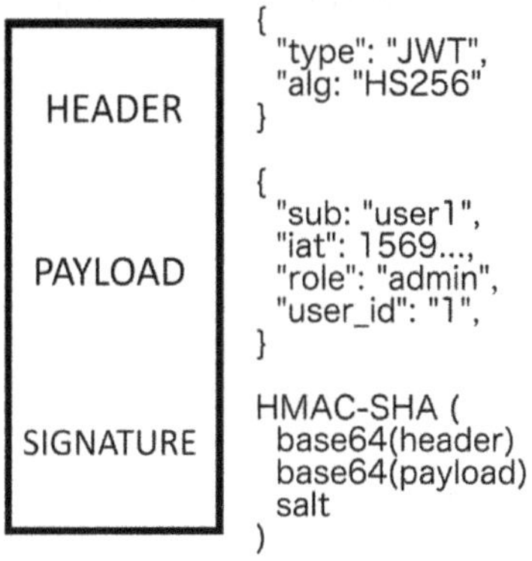

```
                 {
                   "type": "JWT",
                   "alg": "HS256"
   HEADER         }

                 {
                   "sub: "user1",
                   "iat": 1569...,
   PAYLOAD         "role": "admin",
                   "user_id": "1",
                 }

                 HMAC-SHA (
   SIGNATURE       base64(header)
                   base64(payload)
                   salt
                 )
```

Figure 6.1 – JWT structure

As shown in the preceding diagram, a JWT consists of three parts:

- **HEADER**: The header specifies the algorithm that was used to generate the JWT.

- **PAYLOAD**: The payload consists of all the "session" data, which are called claims. The preceding is just a simple representation and does not show the full complexity of a JWT.

- **SIGNATURE**: The signature is calculated from the header and payload. To verify if a JWT has not been tampered with, the signature is compared to the newly generated signature from the actual payload and header.

In our example, we are going to use JWTs, since they are a modern and decentralized method of authentication. Still, you can choose to opt out of this at any point and use regular sessions instead, which can be quickly realized in Express.js and GraphQL.

In the next section, we will look at the different ways of storing the JWT inside the browser and how to transmit between `localStorage` and cookies.

localStorage versus cookies

Let's look at another critical question. It is crucial to understand at least the basics of how authentication works and how it is secured. You are responsible for any faulty implementation that allows data breaches to occur, so always keep this in mind. Where do we store the token we receive from the server?

In whichever direction you send a token, you should always be sure that your communication is secure. For web applications like ours, be sure that HTTPS is enabled and used for all requests. Once the user has successfully authenticated, the client receives the JWT, according to the JWT authentication workflow. A JWT is not tied to any particular storage medium, so you are free to choose whichever you prefer. If we do not store the token when it is received, it will be only available in the memory. While the user is browsing our site, this is fine, but the moment they refresh the page, they will need to log in again because we haven't stored the token anywhere.

There are two standard options: to store the JWT inside `localStorage` or to store it inside a cookie. Let's start by discussing the first option. `localStorage` is the option that's often suggested in tutorials. This is fine, assuming you are writing a single-page web application where the content changes dynamically, depending on the actions of the user and client-side routing. We do not follow any links and load new sites to see new content; instead, the old one is just replaced with the new page that you want to show.

Storing the token in `localStorage` has the following disadvantages:

- `localStorage` is not transmitted on every request. When the page is loaded initially, you are not able to send the token within your request, so resources that need authentication cannot be given back to you. Once your application has finished loading, you must make a second request to your server, including the token to access the secured content. This behavior has the consequence that it is not possible to build server-rendered applications.

- The client needs to implement the mechanics to attach the token to every request that's sent to the server.

- Due to the nature of `localStorage`, there is no built-in expiry date on the client. If, at some point, the token reaches its expiration date, it still exists on the client inside `localStorage`.

- `localStorage` is accessed through pure JavaScript and is therefore open to XSS attacks. If someone manages to integrate custom JavaScript in your code or site through unsanitized inputs, they can read the token from `localStorage`.

There are, however, many advantages of using `localStorage`:

- As `localStorage` is not sent automatically with every request, it is secure against any **Cross-Site-Request-Forgery** (**CSRF**) attacks attempting to run actions from external sites by making random requests.
- `localStorage` is easy to read in JavaScript since it is stored as a key-value pair.
- It supports a bigger data size, which is great for storing an application state or data.

The main problem with storing such critical tokens inside web storage is that you cannot guarantee that there is no unwanted access. Unless you can be sure that every single input is sanitized and you are not relying on any third-party tools that get bundled into your JavaScript code, there is always a potential risk. Just one package you did not build yourself could share your users' web storage with its creator, without you or the user ever noticing. Furthermore, when you are using a public **Content Delivery Network** (**CDN**), the attack base and, consequently, the risk for your application is multiplied.

Now, let's look at cookies. These are great, despite their bad press due to the cookie compliance law that was initiated by the EU. Putting aside the more negative things that cookies can enable the companies to do, such as tracking users, there are still many good things about them. One significant difference compared to `localStorage` is that cookies are sent with every request, including the initial request for the site your application is hosted on.

Cookies come with the following advantages:

- Server-side rendering is no problem at all since cookies are sent with every request.
- No further logic needs to be implemented in the frontend to send the JWT.
- Cookies can be declared as `httpOnly`, which means JavaScript can't access them. It secures our token from XSS attacks.
- Cookies have a built-in expiration date, which can be set to invalidate the cookie in the client browser.
- Cookies can be configured to only be readable from specific domains or paths.
- All browsers support cookies.

These advantages sound good so far, but let's consider the downsides:

- Cookies are generally open to CSRF attacks, which are situations in which an external website makes requests to your API. They expect you to be authenticated and hope that they can execute actions on your behalf. We can't stop the cookie from being sent with each request to your domain. A common prevention tactic is to implement a CSRF token. This special token is also transmitted by your server and saved as a cookie. The external website cannot access the cookie with JavaScript since it is stored under a different domain. Your server does not read a token from the cookies that are transmitted with each request, only from an HTTP header. This behavior guarantees that the token was sent by the JavaScript that was hosted on your application because only this can have access to the token. Setting up the XSRF token for verification, however, introduces a lot of work.

- Accessing and parsing cookies is not intuitive because they are stored as a big comma-separated string.

- They can only store a small amount of data.

So, we can see that both approaches have their advantages and disadvantages.

The most common method is to use `localStorage`, as this is the easiest method. In this book, we will start by using `localStorage`, but later switch over to cookies when using server-side rendering to give you experience with both. You may not need server-side rendering at all. If this is the case, you can skip this part and the cookie implementation too.

In the next section, we are going to implement authentication with GraphQL.

Authentication with GraphQL

The basics of authentication should now be clear to you. Now, our task is to implement a secure way for users to authenticate. If we have a look at our current database, we will see that we are missing the required fields. To do so, follow these steps:

1. Let's prepare and add a `password` field and an `email` field. As we learned in *Chapter 3*, *Connecting to the Database*, we must create a migration to edit our user table. You can look up the commands in that chapter if you have forgotten them:

```
sequelize migration:create --migrations-path src/server/
migrations --name add-email-password-to-post
```

The preceding command generates the new file for us.

2. Replace the content of it and try writing the migration on your own, or you can check for the right commands in the following code snippet:

```
'use strict';

module.exports = {
  up: (queryInterface, Sequelize) => {
    return Promise.all([
      queryInterface.addColumn('Users',
        'email',
        {
          type: Sequelize.STRING,
          unique : true,
        }
      ),
      queryInterface.addColumn('Users',
        'password',
        {
          type: Sequelize.STRING,
        }
      ),
    ]);
  },
  down: (queryInterface, Sequelize) => {
    return Promise.all([
      queryInterface.removeColumn('Users', 'email'),
      queryInterface.removeColumn('Users',
        'password'),
    ]);
  }
};
```

3.   All the fields are simple strings. Execute the migration, as stated in *Chapter 3, Connecting to the Database*. The email address needs to be unique. Our old seed file for the users needs to be updated now to represent the new fields that we have just added. Add the following fields to the first user:

```
password: '$2a$10$bE3ovf9/Tiy/d68bwNUQ0.
zCjwtNFq9ukg9h4rhKiHCb6x5ncKife',
email: 'test1@example.com',
```

Do this for all the users and change the email address for each. Otherwise, it will not work. The password is in hashed format and represents the plain password 123456789. Since we have added new fields in a separate migration, we must add these to the model.

4.   Open and add the following new lines as fields to the user.js file in the model folder:

```
email: DataTypes.STRING,
password: DataTypes.STRING,
```

5.   Now clear the database, run all the migrations, and execute the seeders again.

The first thing we must do is get the login process running. At the moment, we are just faking being logged in as the first user in our database.

Apollo login mutation

In this section, we are going to edit our GraphQL schema and implement the matching resolver function. Follow these steps:

1.   Let's start with the schema and add a new mutation to the RootMutation object of our schema.js file:

```
login (
  email: String!
  password: String!
): Auth
```

The preceding schema gives us a login mutation that accepts an email address and a password. Both are required to identify and authenticate the user. Then, we need to respond with something to the client. For now, the Auth type returns a token, which is a JWT in our case. You might want to add a different option according to your requirements:

```
type Auth {
    token: String
}
```

2. The schema is now ready. Head over to the resolvers file and add the login function inside the mutation object. Before we do this, install and import two new packages:

```
npm install --save jsonwebtoken bcrypt
```

The jsonwebtoken package handles everything that's required to sign, verify, and decode JWTs.

The important part is that all the passwords for our users are not saved as plain text but are first encrypted using hashing, including a random salt. This generated hash cannot be decoded or decrypted as a plain password, but the package can verify if the password that was sent with the login attempt matches the password hash that was saved on the user.

3. Import these packages at the top of the resolvers file:

```
import bcrypt from 'bcrypt';
import JWT from 'jsonwebtoken';
```

4. The login function receives email and password as parameters. It should look as follows:

```
login(root, { email, password }, context) {
  return User.findAll({
    where: {
      email
    },
    raw: true
  }).then(async (users) => {
    if(users.length = 1) {
      const user = users[0];
      const passwordValid = await
```

```
            bcrypt.compare(password, user.password);
        if (!passwordValid) {
          throw new Error('Password does not match');
        }
        const token = JWT.sign({ email, id: user.id },
          JWT_SECRET, {
          expiresIn: '1d'
        });

        return {
          token
        };
      } else {
        throw new Error("User not found");
      }
    });
  },
```

The preceding code goes through the following steps:

I.   We query all the users where the email address matches.

II.   If a user is found, we can move on. It is not possible to have multiple users with the same address, as the MySQL unique constraint forbids this.

III. Next, we use the user's password and compare it with the submitted password, using the `bcrypt` package, as explained previously.

IV.   If the password was correct, we generate a JWT token for the `jwt` variable using the `jwt.sign` function. It takes three arguments: the payload, which is the user ID and their email address; the key that we sign the JWT with; and the amount of time in which the JWT is going to expire.

V.   Finally, we return an object containing our JWT.

> **Note**
>
> Something that you might need to rethink is how much detail you give in an error message. For example, we might not want to distinguish between an incorrect password and a non-existent user. It allows possible attackers or data collectors to know which email address is in use.

The `login` function is not working yet because we are missing `JWT_SECRET`, which is used to sign the JWT. In production, we use the environment variables to pass the JWT secret key into our backend code so that we can use this approach in development too.

5. For Linux or Mac, type the following command directly in the Terminal:

```
export JWT_SECRET=
    awv4BcIzsRysXkhoSAb8t81NENgXSqBruVlLwd45kGdYje
    JHLap9LUJlt9DTdw36DvLcWs3qEkPyCY6vOyNljlh2Er952h2gDzYwG8
    2rslqfTzdVIg89KTaQ4SWI1YGY
```

6. The `export` function sets the `JWT_SECRET` environment variable for you. Replace the JWT provided with a random one. You can use any password generator by setting the character count to 128 and excluding any special characters. Setting the environment variable allows us to read the secret in our application. You must replace it when going to production.

7. Insert the following code at the top of the file:

```
const { JWT_SECRET } = process.env;
```

This code reads the environment variable from the global Node.js `process` object. Be sure to replace the JWT once you publish your application and be sure to always store the secret securely. After letting the server reload, we can send the first login request. We are going to learn how to do this in React later, but the following code shows an example of using Postman:

```
{
    "operationName":null,
    "query": "mutation login($email : String!, $password
        : String!) {
    login(email: $email, password : $password) { token
        }}",
    "variables":{
        "email": "test1@example.com",
        "password": "123456789"
    }
}
```

This request should return a token:

```
{
  "data": {
    "login": {
      "token":"eyJhbGciOiJIUzI1NiIsInR5cCI6IkpXVCJ9.e
        yJlbWFpbCI6InRlc3QxQGV4YW1wbGUuY29tIiwiaWQiOjE
        sImlhdCI6MTUzNzIwNjI0MywiZXhwIjoxNTM3MjkyNjQzf
        Q.HV4dPIBzvU1yn6REMv42N0DS0ZdgebFDXUj0MPHvlY"
    }
  }
}
```

As you can see, we have generated a signed JWT and returned it within the mutation's response. We can continue here and send the token with every request inside the HTTP authorization header. Then, we can get the authentication running for all the other GraphQL queries or mutations that we have implemented so far.

Let's continue and learn how to set up React to work with our authentication on the backend.

The React login form

We need to handle the different authentication states of our application:

- The first scenario is that the user is not logged in and cannot see any posts or chats. In this case, we need to show a login form to allow the user to authenticate themselves.

- The second scenario is that an email and password are sent through the login form. The response needs to be interpreted, and if the result is correct, we need to save the JWT inside `localStorage` of the browser for now.

- When changing `localStorage`, we also need to rerender our React application to show the logged-in state.

- Furthermore, the user should be able to log out again.

- We must also be able to handle if the JWT expires and the user is unable to access any functionalities.

The login form will look as follows:

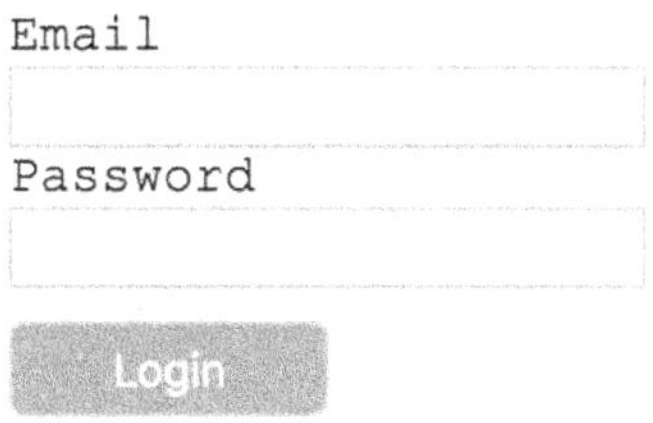

Email

Password

Login

Figure 6.2 – Login form

To get started with the login form, follow these steps:

1. Set up a separate login mutation file inside the apollo folder. It is likely that we only need this component in one place in our code, but it is a good idea to save GraphQL requests in separate files.

2. Build the login form component, which uses the login mutation to send the form data.

3. Create the CurrentUser query to retrieve the logged-in user object.

4. Conditionally render the login form if the user is not authenticated or the real application, such as the newsfeed, if the user is logged in.

5. We will begin by creating a new login.js file inside the mutations folder for the client components:

```
import { gql, useMutation } from '@apollo/client';

export const LOGIN = gql'
  mutation login($email : String!, $password :
    String!) {
    login(email : $email, password : $password) {
      token
    }
  }
';

export const useLoginMutation = () => useMutation(LOGIN);
```

As in the previous mutations, we parse the query string and export the login function from the useMutation Hook.

6. Now, we must implement the actual login form that uses this mutation. To do this, we will create a `loginregister.js` file directly inside the `components` folder. As you may expect, we handle the login and registration of users in one component. Import the dependencies first:

```
import React, { useState } from 'react';
import { useLoginMutation } from '../apollo/mutations/
login';
import Loading from './loading';
import Error from './error';
```

7. The `LoginForm` component will store the form state, display an error message if something goes wrong, show a loading state, and send the login mutation, including the form data. Add the following code beneath the `import` statements:

```
const LoginForm = ({ changeLoginState }) => {
  const [email, setEmail] = useState('');
  const [password, setPassword] = useState('');
  const [login, { loading, error }] =
    useLoginMutation();
  const onSubmit = (event) => {
    event.preventDefault();
    login({
      update(cache, { data: { login } }) {
        if(login.token) {
          localStorage.setItem('jwt', login.token);
          changeLoginState(true);
        }
      }, variables: { email, password }
    });
  }
  return (
    <div className="login">
      {!loading && (
        <form onSubmit={onSubmit}>
          <label>Email</label>
          <input type="text" onChange={(event) =>
            setEmail(event.target.value)} />
```

```
                <label>Password</label>
                <input type="password" onChange={ (event) =>
                    setPassword(event.target.value)} />
                <input type="submit" value="Login" />
            </form>
        )}
        {loading && (<Loading />)}
        {error && (
            <Error><p>There was an error logging in!</p>
            </Error>
        )}
        </div>
    )
}
```

The overall React component is pretty straightforward. We just have one form and two inputs and we store their values in two state variables. The onSubmit function is called when the form is submitted, which will then trigger the login mutation. The update function of the mutation will be a bit different than the other mutations we have had so far. We don't write the return value in the Apollo cache; instead, we store the JWT inside localStorage. The syntax is pretty simple. You can directly use localStorage.get and localStorage.set to interact with the web storage.

After saving the JWT to localStorage, we call a changeLoginState function, which we will implement in the next step. The idea of this function is to have one global switch to change a user from logged in to logged out or vice versa.

8. We now need to export a component that will be used by our application. The easiest thing to do is set up a wrapper component that handles the login and sign-up cases for us.

Insert the following code for the wrapper component:

```
const LoginRegisterForm = ({ changeLoginState }) => {
    return (
        <div className="authModal">
            <div>
                <LoginForm changeLoginState={changeLoginState}
                />
            </div>
```

```
        </div>
      )
    }
```

```
    export default LoginRegisterForm
```

This component just renders the login form and passes the `changeLoginState` function.

All the basics for authenticating the user are now ready, but they have not been imported yet or displayed anywhere. Open the `App.js` file. There, we will directly display the feed, chats, and the top bar. The user should not be allowed to see everything if they are not logged in. Continue reading to change this.

9. Import the new form that we have just created and import the `useEffect` Hook from React:

    ```
    import LoginRegisterForm from './components/
    loginregister';
    ```

10. Now, we must store whether the user is logged in or not and, also on the first render of our application, check the login state based on `localStorage`. Add the following code to the `App` component:

    ```
    const [loggedIn, setLoggedIn] = useState(!!localStorage.
    getItem('jwt'));
    ```

    When loading our page, we have the `loggedIn` state variable to store the current logged-in status inside. The default value is either `true` if the token exists or `false` if not.

11. Then, in the `return` statement, we can use conditional rendering to show the login form when the `loggedIn` state variable is set to `false`, which means that there is no JWT inside our `localStorage`:

    ```
    {loggedIn && (
      <div>
        <Bar changeLoginState={setLoggedIn} />
        <Feed />
        <Chats />
      </div>
    )}
    ```

```
{!loggedIn && <LoginRegisterForm
changeLoginState={setLoggedIn} />}
```

As you can see, we pass the `setLoggedIn` function to the login form, which is then able to trigger a logged-in state so that React can rerender and show the logged-in area. We call this property `changeLoginState` and use it inside the login form inside the `update` method in the login mutation.

12. Add the CSS from the official GitHub repository:

    `https://github.com/PacktPublishing/Full-Stack-Web-Development-with-GraphQL-and-React-Second-Edition`

Once we've logged in, our application will present us with the common posts feed, as it did previously. The authentication flow is now working, but there is one more open task. In the next section, we will allow new users to register at Graphbook.

Apollo signup mutation

You should now be familiar with creating new mutations. To do so, follow these steps:

1. First, edit the schema to accept the new mutation:

    ```
    signup (
      username: String!
      email: String!
      password: String!
    ): Auth
    ```

    We only need the `username`, `email`, and `password` properties, which were mentioned in the preceding code, to accept new users. If your application requires a gender or something else, you can add it here. When we're trying to sign up, we need to ensure that neither the email address nor the username has already been taken.

2. Copy over the code to implement the resolver for signing up new users:

    ```
    signup(root, { email, password, username }, context) {
      return User.findAll({
        where: {
          [Op.or]: [{email}, {username}]
        },
        raw: true,
      }).then(async (users) => {
    ```

```
    if(users.length) {
      throw new Error('User already exists');
    } else {
      return bcrypt.hash(password, 10).then((hash) => {
        return User.create({
          email,
          password: hash,
          username,
          activated: 1,
        }).then((newUser) => {
          const token = JWT.sign({ email, id:
            newUser.id }, JWT_SECRET,
          {
            expiresIn: '1d'
          });
          return {
            token
          };
        });
      });
    }
  });
},
```

Let's go through this code step by step:

I.   As we mentioned previously, first, we must check if a user with the same email
      or username exists. If this is the case, we throw an error. We use the Op.or
      Sequelize operator to implement the MySQL OR condition.

II.   If the user does not exist, we can hash the password using bcrypt. You
      cannot save the plain password for security reasons. When running the
      bcrypt.hash function, a random salt is used to make sure nobody ever gets
      access to the original password. This command takes quite some computing
      time, so the bcrypt.hash function is asynchronous, and the promise must be
      resolved before we continue.

III. The encrypted password, including the other data the user has sent, is then inserted into our database as a new user.

IV. After creating the user, we generate a JWT and return it to the client. The JWT allows us to log the user in directly once they've signed up. If you do not want this behavior, you can just return a message to indicate that the user has signed up successfully.

Now, you can test the `signup` mutation again with Postman while starting the backend using `npm run server`. With that, we have finished the backend implementation. So, let's start working on the frontend.

React signup form

The registration form is nothing special. We will follow the same steps that we took with the login form:

1. Clone the `LoginMutation` component, replace the request at the top with the `signup` mutation, and hand over the `signup` method to the underlying children.

2. At the top, import all the dependencies and then parse the new query:

```
import { gql, useMutation } from '@apollo/client';


export const SIGNUP = gql'
  mutation signup($email : String!, $password :
    String!, $username : String!) {
    signup(email : $email, password : $password,
      username : $username) {
      token
    }
  }
';


export const useSignupMutation = () =>
useMutation(SIGNUP);
```

As you can see, the `username` field is new here, which we send with every `signup` request. The logic itself has not changed, so we still extract the JWT from the `signup` field when logging the user in after a successful request.

It's good to see that the `login` and `signup` mutations are quite similar. The biggest change is that we conditionally render the login form or the registration form. Follow these steps:

1. Import the new mutation into the `loginregister.js` file:

    ```
    import { useSignupMutation } from '../apollo/mutations/
    signup';
    ```

2. Then, replace the complete `LoginRegisterForm` component with the following new one:

    ```
    const LoginRegisterForm = ({ changeLoginState }) => {
      const [showLogin, setShowLogin] = useState(true);

      return (
        <div className="authModal">
          {showLogin && (
            <div>
              <LoginForm
                changeLoginState={changeLoginState} />
              <a onClick={() => setShowLogin(false)}>
                Want to sign up? Click here</a>
            </div>
          )}
          {!showLogin && (
            <div>
              <RegisterForm
                changeLoginState={changeLoginState} />
              <a onClick={() => setShowLogin(true)}>
                Want to login? Click here</a>
            </div>
          )}
        </div>
      )
    }
    ```

You should notice that we are storing a `showLogin` variable in the component state. This decides if the login or register component is shown, which handles the actual business logic.

3.   Then, add a separate component for the register form before the export statement:

```
const RegisterForm = ({ changeLoginState }) => {
  const [email, setEmail] = useState('');
  const [password, setPassword] = useState('');
  const [username, setUsername] = useState('');
  const [signup, { loading, error }] =
    useSignupMutation();
  const onSubmit = (event) => {
    event.preventDefault();
    signup({
      update(cache, { data: { login } }) {
        if(login.token) {
          localStorage.setItem('jwt', login.token);
          changeLoginState(true);
        }
      }, variables: { email, password, username }
    });
  }
  return (
    <div className="login">
      {!loading && (
        <form onSubmit={onSubmit}>
          <label>Email</label>
          <input type="text" onChange={ (event) =>
            setEmail(event.target.value)} />
          <label>Username</label>
          <input type="text" onChange={ (event) =>
            setUsername(event.target.value)} />
          <label>Password</label>
          <input type="password" onChange={(event) =>
            setPassword(event.target.value)} />
          <input type="submit" value="Sign up" />
        </form>
```

```
        ) }
        {loading && (<Loading />)}
        {error && (
          <Error><p>There was an error logging in!</p>
          </Error>
        ) }
      </div>
    )
  }
```

In the preceding code, I added the `username` field, which must be given to the mutation. Everything is now set to invite new users to join our social network and log in as often as they want.

In the next section, we will learn how to use authentication with our existing GraphQL requests.

Authenticating GraphQL requests

The problem is that we are not using authentication everywhere at the moment. We are verifying that the user is who they say they are, but we are not rechecking this when the requests for chats or messages come in. To accomplish this, we must send the JWT token, which we generated specifically for this case, with every Apollo request. On the backend, we must specify which request requires authentication, read the JWT from the HTTP authorization header, and verify it. Follow these steps:

1. Open the `index.js` file from the `apollo` folder for the client-side code. Our `ApolloClient` is currently configured as explained in *Chapter 4, Hooking Apollo into React*. Before we send any requests, we must read the JWT from `localStorage` and add it as an HTTP authorization header. Inside the `link` property, we have specified the links for our `ApolloClient` processes. Before we configure the HTTP link, we must insert a third preprocessing Hook, as follows:

```
const AuthLink = (operation, next) => {
  const token = localStorage.getItem('jwt');
  if(token) {
    operation.setContext(context => ({
      ...context,
      headers: {
        ...context.headers,
        Authorization: 'Bearer ${token}',
```

```
        },
    }));
  }
  return next(operation);
};
```

Here, we have called the new link AuthLink because it allows us to authenticate the client on the server. You can copy the AuthLink approach for other situations where you need to customize the header of your Apollo requests. Here, we just read the JWT from localStorage and, if it is found, we construct the header using the spread operator and add our token to the Authorization field as a Bearer Token. This is everything that needs to be done on the client side.

2. To clarify things, take a look at the following link property to learn how to use this new preprocessor. No initialization is required; it is merely a function that is called every time a request is made. Copy the link configuration to our Apollo Client setup:

```
link: from([
  onError(({ graphQLErrors, networkError }) => {
    if (graphQLErrors) {
      graphQLErrors.map(({ message, locations, path })
        =>
      console.log('[GraphQL error]: Message:
        ${message}, Location:
      ${locations}, Path: ${path}'));
      if (networkError) {
        console.log('[Network error]:
          ${networkError}');
      }
    }
  }),
  AuthLink,
  new HttpLink({
    uri: 'http://localhost:8000/graphql',
    credentials: 'same-origin',
  }),
]),
```

3. Let's install one dependency that we require:

    ```
    npm install --save @graphql-tools/utils
    ```

4. For our backend, we need a pretty complex solution. Create a new file called
    auth.js inside the GraphQL services folder. We want to be able to mark specific
    GraphQL requests in our schema with a so-called directive. If we add this directive
    to our GraphQL schema, we can execute a function whenever the marked GraphQL
    action is requested. In this function, we can verify whether the user is logged in or
    not. Have a look at the following function and save it in the auth.js file:

    ```
    import { mapSchema, getDirective, MapperKind } from '@
    graphql-tools/utils';

    function authDirective(directiveName) {
      const typeDirectiveArgumentMaps = {};

      return {
        authDirectiveTypeDefs: 'directive
          @${directiveName} on QUERY | FIELD_DEFINITION |
            FIELD',
        authDirectiveTransformer: (schema) =>
          mapSchema(schema, {
          [MapperKind.TYPE]: (type) => {
            const authDirective = getDirective(schema,
              type, directiveName)?.[0];
            if (authDirective) {
              typeDirectiveArgumentMaps[type.name] =
                authDirective;
            }
            return undefined;
          },
          [MapperKind.OBJECT_FIELD]: (fieldConfig,
            _fieldName, typeName) => {
            const authDirective = getDirective(schema,
              fieldConfig, directiveName)?.[0] ??
                typeDirectiveArgumentMaps[typeName];
            if (authDirective) {
    ```

```
            const { resolve = defaultFieldResolver } =
                fieldConfig;
            fieldConfig.resolve = function (source,
                args, context, info) {
                if (context.user) {
                    return resolve(source, args, context,
                        info);
                }
                throw new Error("You need to be logged
                                    in.");
            }
            return fieldConfig;
        }
      }
    }),
  };
}

export default authDirective;
```

Starting from the top, we import three things from the @graphql/utils package:

A. The mapSchema function takes two arguments. The first is the actual GraphQL schema and then an object of functions that can transform the schema.

B. The getDirective function will read the schema and try to get the specified directiveName. Based on that, we can do anything that we want.

C. MapperKind is just a set of types that we can use. We are using that to only run functions for specific types.

This function or directive will read the user from the context and pass it to our resolvers where the directive is specified within our GraphQL schema.

5. We must load the new authDirective function in the graphql index.js file, which sets up the whole Apollo Server:

```
import authDirective from './auth';
```

6. Before we create our executable schema, we must extract the new schema transformer from the `authDirective` function. After creating the executable schema, we must pass it to the transformer so that `authDirective` starts to work. Replace the current schema creation with the following code:

```
const { authDirectiveTypeDefs, authDirectiveTransformer }
= authDirective('auth');
let executableSchema = makeExecutableSchema({
    typeDefs: [authDirectiveTypeDefs, Schema],
    resolvers: Resolvers.call(utils),
});
executableSchema =
authDirectiveTransformer(executableSchema);
```

7. To verify what we have just done, go to the GraphQL schema and edit `postsFeed` `RootQuery` by adding `@auth` to the end of the line, like this:

```
postsFeed(page: Int, limit: Int): PostFeed @auth
```

8. Because we are using a new directive, we also must define it in our GraphQL schema so that our server knows about it. Copy the following code directly to the top of the schema:

```
directive @auth on QUERY | FIELD_DEFINITION | FIELD
```

This tiny snippet tells Apollo Server that the `@auth` directive can be used with queries, fields, and field definitions so that we can use it everywhere.

If you reload the page and manually set the `loggedIn` state variable to true via React Developer Tools, you will see the following error message:

```
GraphQL error: You need to be logged in.
```

Figure 6.3 – GraphQL login error

Since we implemented the error component earlier, we are now correctly receiving an unauthenticated error for the `postsFeed` query if the user is not logged in. How can we use the JWT to identify the user and add it to the request context?

> **Note**
>
> Schema directives are a complex topic as there are many important things to bear in mind regarding Apollo and GraphQL. I recommend that you read up on directives in detail in the official Apollo documentation: `https://www.graphql-tools.com/docs/introduction`.

In *Chapter 2, Setting Up GraphQL with Express.js*, we set up Apollo Server by providing the executable schema and the context, which has been the request object until now. We must check if the JWT is inside the request. If this is the case, we need to verify it and query the user to see if the token is valid. Let's start by verifying the authorization header. Before doing so, import the new dependencies into the GraphQL `index.js` file:

```
import JWT from 'jsonwebtoken';
const { JWT_SECRET } = process.env;
```

The `context` field of the `ApolloServer` initialization must look as follows:

```
context: async ({ req }) => {
  const authorization = req.headers.authorization;
  if(typeof authorization !== typeof undefined) {
      var search = "Bearer";
      var regEx = new RegExp(search, "ig");
      const token = authorization.replace(regEx,
          '').trim();
      return JWT.verify(token, JWT_SECRET, function(err,
          result) {
          if(err) {
              return req;
          } else {
              return utils.db.models.User.findByPk(
                  result.id).then((user) => {
                      return Object.assign({}, req, { user });
                  });
          }
      });
  } else {
      return req;
  }
},
```

Here, we have extended the `context` property of the `ApolloServer` class to a full-featured function. We read the `auth` token from the headers of the requests. If the `auth` token exists, we need to strip out the bearer string, because it is not part of the original token that was created by our backend. The Bearer Token is the best method of JWT authentication.

> **Note**
>
> There are other authentication methods available, such as basic authentication, but the bearer method is the best to follow. You can find a detailed explanation under RFC6750 by the IETF at `https://tools.ietf.org/html/rfc6750`.

Afterward, we must use the `JWT.verify` function to check if the token matches the signature that's been generated by the secret from the environment variables. The next step is to retrieve the user once they've been verified successfully. Replace the content of the `verify` callback with the following code:

```
if(err) {
    return req;
} else {
    return utils.db.models.User.findByPk(result.id).then((
        user) => {
        return Object.assign({}, req, { user });
    });
}
```

If the `err` object in the previous code has been filled, we can only return the ordinary request object, which triggers an error when it reaches the `auth` directive, since there is no user attached. If there are no errors, we can use the `utils` object we are already passing to the Apollo Server setup to access the database. If you need a reminder, take a look at *Chapter 2, Setting Up GraphQL with Express.js*. After querying the user, we must add them to the request object and return the merged user and request object as the context. This leads to a successful response from our authorizing directive.

Now, let's test this behavior. Start the frontend with `npm run client` and the backend using `npm run server`. Don't forget that all Postman requests now have to include a valid JWT if the `auth` directive is used in the GraphQL query. You can run the login mutation and copy it over to the authorization header to run any query. We are now able to mark any query or mutation with the authorization flag and, as a result, require the user to be logged in.

Accessing the user context from resolver functions

At the moment, all the API functions of our GraphQL server allow us to simulate the user by selecting the first that's available from the database. As we have just introduced a full-fledged authentication, we can now access the user from the request context. This section quickly explains how to do this for the chat and message entities. We will also implement a new query called currentUser, where we retrieve the logged-in user in our client.

Chats and messages

First of all, you must add the @auth directive to the chats inside GraphQL's RootQuery to ensure that users need to be logged in to access any chats or messages.

Take a look at the resolver function for the chats. Currently, we are using the findAll method to get all users, take the first one, and query for all the user's chats. Replace this code with the following new resolver function:

```
chats(root, args, context) {
  return Chat.findAll({
    include: [{
      model: User,
      required: true,
      through: { where: { userId: context.user.id } },
    },
    {
      model: Message,
    }],
  });
},
```

Here, we don't retrieve the user; instead, we directly insert the user ID from the context, as shown in the preceding code. That's all we have to do: all the chats and messages that belong to the logged-in user are queried directly from the chats table.

We would need to copy this for the mutations for chats and messages and all the other queries and mutations that we have at the moment.

CurrentUser GraphQL query

JWTs allow us to query for the currently logged-in user. Then, we can display the correct authenticated user in the top bar. To request the logged-in user, we require a new query called `currentUser` on our backend. In the schema, you simply have to add the following line to the `RootQuery` queries:

```
currentUser: User @auth
```

Like the `postsFeed` and `chats` queries, we also need the `@auth` directive to extract the user from the request context.

Similarly, in the resolver functions, you only need to insert the following three lines:

```
currentUser(root, args, context) {
    return context.user;
},
```

We return the user from the context right away, because it is already a user model instance with all the appropriate data being returned by Sequelize. On the client side, we create this query in a separate component and file. Bear in mind that you don't need to pass the result to all the children because this is done automatically by `ApolloConsumer` later. You can follow the previous query component examples to see this. Just create a file called `currentUserQuery.js` in the `queries` folder with the following content:

```
import { gql, useQuery } from '@apollo/client';

export const GET_CURRENT_USER = gql'
    query currentUser {
        currentUser {
            id
            username
            avatar
        }
    }
';

export const useCurrentUserQuery = (options) => useQuery(GET_
CURRENT_USER, options);
```

Now, you can import the new query inside the `App.js` file and add the following line to the `App` component:

```
const { data, error, loading, refetch } =
useCurrentUserQuery();
if(loading) {
    return <Loading />;
}
```

Here, we executed the `useCurrentUserQuery` Hook to ensure that the query has been executed at a global level for all the components. Also, we show a loading indicator until the request has finished to ensure that the user is loaded before we do anything else.

Every time the `loggedIn` state variable is `true`, we render the components. To get access to the response, we must use `ApolloConsumer` in the bar component, which we implemented in the previous chapter. We run the `currentUser` query in the `App.js` file to ensure that all the child components can rely on the Apollo cache to access the user before being rendered.

Instead of having a hardcoded fake user inside `ApolloConsumer`, we can use the the `client.readQuery` function to extract the data stored in the `ApolloClient` cache to give it to the underlying child component. Replace the current consumer with the following code:

```
import React from 'react';
import { ApolloConsumer } from '@apollo/client';
import { GET_CURRENT_USER } from '../../apollo/queries/
currentUserQuery';


export const UserConsumer = ({ children }) => {
  return (
    <ApolloConsumer>
      {client => {
        const result = client.readQuery({ query:
          GET_CURRENT_USER });
        return React.Children.map(children,
          function(child){
          return React.cloneElement(child, { user:
            result?.currentUser ? result.currentUser : null
            });
```

```
        });
      }}
    </ApolloConsumer>
  )
}
```

Here, we passed the extracted `currentUser` result from the `client.readQuery` method to all the wrapped children of the current component.

The chats that are displayed from now on, as well as the user in the top bar, are no longer faked; instead, they are filled with the data related to the logged-in user.

The mutations to create new posts or messages still use a static user ID. We can switch over to the real logged-in user in the same way as we did previously in this section by using the user ID from the `context.user` object. You should now be able to do this on your own.

Logging out using React

To complete the circle, we still have to implement the functionality to log out. There are two cases when the user can be logged out:

- The user wants to log out and hits the logout button.
- The JWT has expired after 1 day as specified; the user is no longer authenticated, and we have to set the state to logged out.

Follow these steps to accomplish this:

1. We will begin by adding a new logout button to the top bar of our application's frontend. To do this, create a new `logout.js` component inside the `bar` folder. It should look as follows:

```
import React from 'react';
import { withApollo } from '@apollo/client/react/hoc';

const Logout = ({ changeLoginState, client }) => {
  const logout = () => {
    localStorage.removeItem('jwt');
    changeLoginState(false);
    client.stop();
    client.resetStore();
```

```
    }

    return (
        <button className="logout" onClick={logout}>Logout
        </button>
    );
}
```

```
export default withApollo(Logout);
```

As you can see, the logout button triggers the component's logout method when it is clicked. Inside the `logout` method, we remove the JWT from `localStorage` and execute the `changeLoginState` function that we receive from the parent component. Be aware that we do not send a request to our server to log out; instead, we remove the token from the client. This is because there is no black or white list that we are using to disallow or allow a certain JWT to authenticate on our server. The easiest way to log out a user is to remove the token on the client side so that neither the server nor the client has it.

We also reset the client cache. When a user logs out, we must remove all data. Otherwise, other users on the same browser will be able to extract all the data, which we must prevent. To gain access to the underlying Apollo Client, we must import the `withApollo` **HOC** (short for **Higher-Order-Component**) and export the `Logout` component wrapped inside it. When logging out, we must execute the `client.stop` and `client.resetStore` functions so that all the data is deleted.

2. To use our new `Logout` component, open the `index.js` file from the `bar` folder and import it at the top. We can render it within the `div` top bar, below the other inner `div` tag:

```
<div className="buttons">
  <Logout changeLoginState={changeLoginState}/>
</div>
```

Here, we pass the `changeLoginState` function to the `Logout` component.

3. Extract the `changeLoginState` function from the `Bar` component props, as follows:

```
const Bar = ({ changeLoginState }) => {
```

4. In the App.js file, you must implement one more function to handle the current user query correctly. If we are not logged in and then log in, we need to fetch the current user. And if we log out, we need to either set or be able to easily fetch the current user query again. Add the following function:

```
const handleLogin = (status) => {
    refetch().then(() => {
        setLoggedIn(status);
    }).catch(() => {
        setLoggedIn(status);
    });
}
```

5. Hand this function over not only to LoginRegisterForm but also to the Bar component, as follows:

```
<Bar changeLoginState={handleLogin} />
```

6. If you copy the complete CSS from the official GitHub repository, you should see a new button at the top-right corner of the screen when you are logged in. Hitting it logs you out and requires you to sign in again since the JWT has been deleted.

7. The other situation in which we implement logout functionality is when the JWT we are using expires. In this case, we log the user out automatically and require them to log in again. Go to the App component and add the following lines:

```
useEffect(() => {
    const unsubscribe = client.onClearStore(
        () => {
            if (loggedIn) {
                setLoggedIn(false)
            }
        }
    );
    return () => {
        unsubscribe();
    }
}, []);
```

Here, we are using the client.onClearStore event, which is caught through the client.onClearStore function once the client store is cleared.

8. To get the preceding code to work, we must access the Apollo Client in our App component. The easiest way to do this is to use the withApollo HoC. Just import it from the @apollo/client package in the App.js file:

```
import { withApollo } from '@apollo/client/react/hoc';
```

9. Then, export the App component – not directly, but through the HoC – and extract the client property. The following code must go directly beneath the App component:

```
export default withApollo(App);
```

Now, the component can access the client through its properties. The clearStore event is thrown whenever the client restore is reset, as its name suggests. You are going to see why we need this shortly. When listening to events in React, we have to stop listening when the component is unmounted. We handle this inside the useEffect Hook in the preceding code. Now, we must reset the client store to initiate the logout state. When the event is caught, we execute the changeLoginState function automatically. Consequently, we could remove the section that we passed changeLoginState to the logout button initially because it is no longer needed, but this is not what we want to do here.

10. Extract the client from the App component props, like so:

```
const App = ({ client }) => {
```

11. Go to the index.js file in the apollo folder. There, we already caught and looped over all the errors that were returned from our GraphQL API. What we must do now is loop over all the errors but check each of them for an UNAUTHENTICATED error. Then, we must execute the client.clearStore function. Insert the following code into the Apollo Client setup:

```
onError(({ graphQLErrors, networkError }) => {
  if (graphQLErrors) {
    graphQLErrors.map(({ message, locations, path,
        extensions }) => {
      if(extensions.code === 'UNAUTHENTICATED') {
        localStorage.removeItem('jwt');
        client.clearStore()
      }
      console.log('[GraphQL error]: Message:
          ${message}, Location:
```

```
        ${locations}, Path: ${path}');
      });
    if (networkError) {
      console.log('[Network error]: ${networkError}');
    }
  }
}),
```

As you can see, we access the `extensions` property of the error. The `extensions.code` field holds the specific error type that's returned. If we are not logged in, we remove the JWT and then reset the store. By doing this, we trigger the event in our `App` component, which sends the user back to the login form.

A further extension would be to offer a refresh token API function. This feature could be run every time we successfully use the API. The problem with this is that the user would stay logged in forever, so long as they are using the application. Usually, this is not a problem, but if someone else is accessing the same computer, they will be authenticated as the original user. There are different ways to implement these kinds of functionalities to make the user experience more comfortable, but I am not a big fan of these for security reasons.

Summary

Until now, one of the main issues we had with our application is that we didn't have any authentication. We can now tell who is logged in every time a user accesses our application. This allows us to secure the GraphQL API and insert new posts or messages into the name of the correct user. In this chapter, we discussed the fundamental aspects of JWTs, `localStorage`, and cookies. We also looked at how hashed password verification and signed tokens work. We then covered how to implement JWTs inside React and how to trigger the correct events to log in and log out.

In the next chapter, we are going to implement image uploads with a reusable component that allows the user to upload new avatar images.

7

Handling Image Uploads

All social networks have one thing in common: each of them allows its users to upload custom and personal pictures, videos, or any other kind of document. This feature can take place inside chats, posts, groups, or profiles. To offer the same functionality, we are going to implement an image upload feature in Graphbook.

This chapter will cover the following topics:

- Setting up Amazon Web Services
- Configuring an AWS S3 bucket
- Accepting file uploads on the server
- Uploading images with React through Apollo
- Cropping images

Technical requirements

The source code for this chapter is available in the following GitHub repository:

`https://github.com/PacktPublishing/Full-Stack-Web-Development-with-GraphQL-and-React-Second-Edition/tree/main/Chapter07`

Setting up Amazon Web Services

First, I have to mention that Amazon—or, to be specific, **Amazon Web Services (AWS)**—is not the only provider of hosting, storage, or computing systems. There are many such providers, including the following:

- Heroku
- DigitalOcean
- Google Cloud
- Microsoft Azure

AWS offers everything that you need to run a full-fledged web application, as with all the other providers. Furthermore, it is also widely used, which is why we are focusing on AWS for this book.

Its services span from databases to object storage, to security services, and so much more. Besides, AWS is the go-to solution that you will find in most other books and tutorials, and many big companies use it to power their complete infrastructure.

This book uses AWS for serving static files, such as images, to run the production database and the Docker container for our application.

Before continuing with this chapter, you will be required to have an account for AWS. You can create one on the official web page at `https://aws.amazon.com/`. For this, you will need a valid credit card; you can also run nearly all of the services on the Free Tier while working through this book without facing any problems.

Once you have successfully registered for AWS, you will see the following dashboard. This screen is called the **AWS Management Console**:

Figure 7.1 – AWS Management Console

The next section will cover the options for storing files with AWS.

Configuring an AWS S3 bucket

For this chapter, we will require a storage service to save all uploaded images. AWS provides different storage types for various use cases. In our scenario of a social network, we will have dozens of people accessing many images at once. **AWS Simple Storage Service (S3)** is the best option for our scenario. Follow these steps to set up an S3 bucket:

1. You can visit the **Amazon S3** screen by clicking on the **Services** drop-down menu at the top of the page, and then looking under the **Storage** category in the drop-down menu. There, you will find a link to S3. Having clicked on it, the screen will look like this:

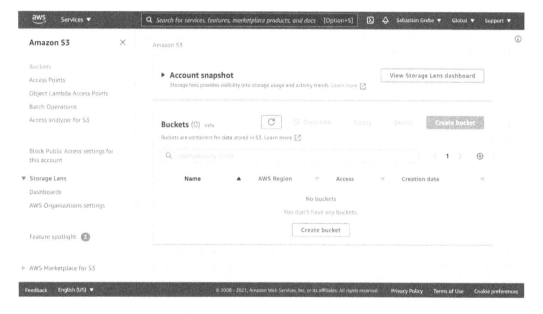

Figure 7.2 – S3 management screen

In S3, you create a bucket inside of a specific AWS region, where you can store files.

The preceding screen provides many features for interacting with your S3 bucket. You can browse all of the files, upload your files via the management interface, and configure more settings.

2. We will now create a new bucket for our project by clicking on **Create bucket** in the upper-right corner, as shown in *Figure 7.2*. You will be presented with a form, as shown in the following screenshot. To create a bucket, you must fill it out:

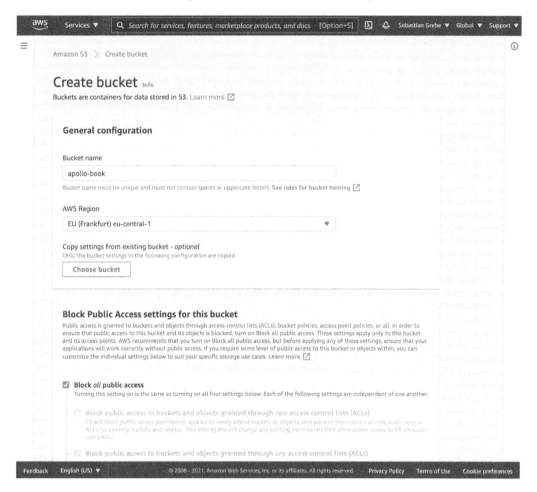

Figure 7.3 – S3 bucket wizard

The bucket has to have a unique name across all buckets in S3. Then, we need to pick a region. For me, **EU (Frankfurt) eu-central-1** is the best choice, as it is the nearest origin point. Choose the best option for you, since the performance of a bucket corresponds to the distance between the region of the bucket and its accessor.

Then, you need to uncheck the **Block all public access** option and also check the acknowledgment with the warning sign. AWS shows us this warning because we should only give public access to S3 buckets when really required. It should look like this:

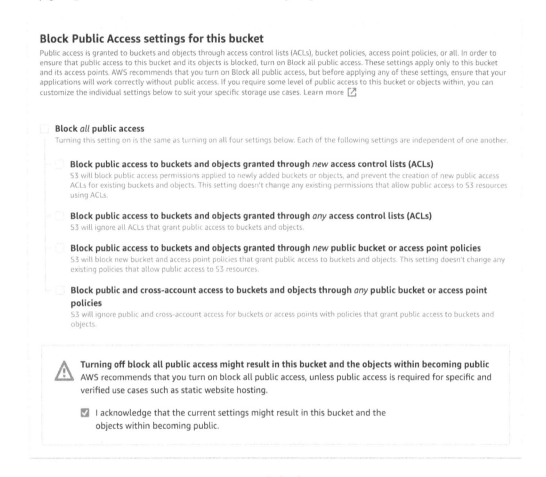

Figure 7.4 – S3 bucket access

For our use case, we can stay with the default settings provided for all the other options in this form wizard. The other options can be helpful in more advanced scenarios. AWS offers many features, such as a complete access log and versioning.

> **Note**
>
> Many bigger companies have users across the globe, which requires a highly available application. When you reach this point, you can create many more S3 buckets in other regions, and you can set up the replication of one bucket to others living in various regions around the world. The correct bucket can then be distributed with AWS CloudFront and a router specific to each user. This approach gives every user the best possible experience.

Finish the setup process by clicking on **Create bucket** at the bottom of the page. You will be redirected back to the table view for all buckets.

Generating AWS access keys

Before implementing the upload feature, we must create an AWS **application programming interface (API)** key to authorize our backend at AWS, in order to upload new files to the S3 bucket.

Click on your username in the top bar of the AWS management screen. There, you will find a tab called **My Security Credentials**, which navigates to a screen offering various options to secure access to your AWS account.

You will be confronted with a dialog box like this:

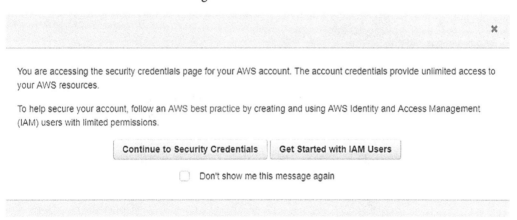

Figure 7.5 – S3 Identity and Access Management (IAM) dialog

You can click on **Continue to Security Credentials** to continue. It is generally recommended to use AWS IAM, which allows you to efficiently manage secure access to AWS resources with separate IAM users. Throughout this book, we are going to use the root user in the same way that we are now, but I recommend looking at AWS IAM when writing your next application.

You should now see the credentials page, with a big list of different methods for storing credentials. This is how it should look:

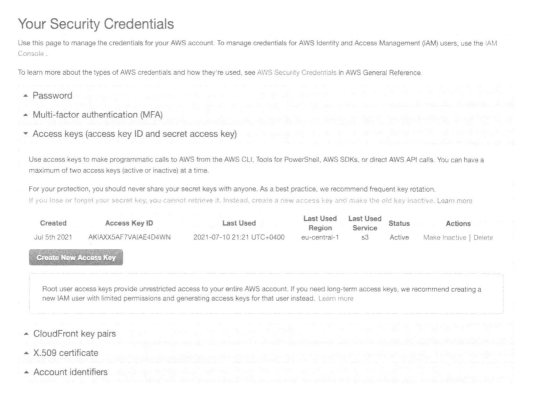

Figure 7.6 – AWS access keys

In the list, expand the tab titled **Access keys (access key ID and secret access key)** shown in the preceding screenshot. In this tab, you will find all access tokens for your AWS account.

To generate a new access token, click on **Create New Access Key**. The output should look like this:

Figure 7.7 – AWS access key

The best practice is to download the key file as prompted and save it somewhere securely, just in case you lose the key at any time. You cannot retrieve access keys again after closing the window, so if you lose them, you will have to delete the old key and generate a new one.

> **Note**
>
> This approach is acceptable for explaining the basics of AWS. With such a huge platform, there are further steps that you have to take to secure your application even more. For example, it is recommended to renew API keys every 90 days. You can read more about all of the best practices at `https://docs.aws.amazon.com/general/latest/gr/aws-access-keys-best-practices.html`.

As you can see in *Figure 7.7*, AWS gives us two tokens. Both are required to gain access to our S3 bucket.

Now, we can start to program the uploading mechanism.

Uploading images to Amazon S3

Implementing file uploads and storing files is always a huge task, especially for image uploads in which the user may want to edit their files again.

For our frontend, the user should be able to drag and drop their image into a dropzone, crop the image, and then submit it when they are finished. The backend needs to accept file uploads in general, which is not easy at all. The files must be processed and then stored efficiently so that all users can access them quickly.

As this is a vast topic, the chapter only covers the basic upload of images from React, using a multipart **HyperText Transfer Protocol** (**HTTP**) POST request to our GraphQL API, and then transferring the image to our S3 bucket. When it comes to compressing, converting, and cropping, you should check out further tutorials or books on this topic, including techniques for implementing them in the frontend and backend, since there is a lot to think about. For example, in many applications, it makes sense to store images in various resolutions that will be shown to the users in different situations, in order to save bandwidth.

Let's start by implementing the upload process on the backend.

GraphQL image upload mutation

When uploading images to S3, it is required to use an API key, which we have already generated. Because of this, we cannot directly upload the files from the client to S3 with the API key. Anyone accessing our application could read out the API key from the JavaScript code and access our bucket without us knowing.

Uploading images directly from the client into the bucket is generally possible, however. To do this, you would need to send the name and type of the file to the server, which would then generate a **Uniform Resource Locator** (**URL**) and signature. The client can then use the signature to upload the image. This technique results in many round trips for the client and does not allow us to postprocess the image, such as by converting or compressing, if needed.

A better solution is to upload the images to our server, have the GraphQL API accept the file, and then make another request to S3—including the API key—to store the file in our bucket.

We have to prepare our backend to communicate with AWS and accept file uploads. The preparation steps are listed here:

1. We install the official npm package to interact with AWS. It provides everything that's needed to use any AWS feature, not just S3. Also, we install graphql-upload, which provides some tools to resolve the file from any GraphQL request. The code to do this is illustrated here:

    ```
    npm install --save aws-sdk graphql-upload
    ```

2. Inside the server index.js file, we need to add the initialization of the graphql-upload package. For that, import the Express dependency at the top, as follows:

    ```
    import { graphqlUploadExpress } from 'graphql-upload';
    ```

3. Inside the `graphql` case at the end of the file, before executing the `applyMiddleware` function, we need to initiate it, as follows:

```
case 'graphql':
  (async () => {
    await services[name].start();
    app.use(graphqlUploadExpress());
    services[name].applyMiddleware({ app });
  })();
  break;
```

4. The next thing to do is edit the GraphQL schema and add an `Upload` scalar to the top of it. The scalar is used to resolve details such as the **Multipurpose Internet Mail Extensions** (**MIME**) type and encoding when uploading files. Here's the code you'll need:

```
scalar Upload
```

5. Add the `File` type to the schema. This type returns the filename and the resulting URL under which the image can be accessed in the browser. The code is illustrated in the following snippet:

```
type File {
  filename: String!
  mimetype: String!
  encoding: String!
  url: String!
}
```

6. Create a new `uploadAvatar` mutation. The user needs to be logged in to upload avatar images, so append the `@auth` directive to the mutation. The mutation takes the previously mentioned `Upload` scalar as input. The code is illustrated in the following snippet:

```
uploadAvatar (
  file: Upload!
): File @auth
```

7. Next, we will implement the mutation's resolver function in the `resolvers.js` file. For this, we will import and set up our dependencies at the top of the `resolvers.js` file, as follows:

```
import { GraphQLUpload } from 'graphql-upload';
import aws from 'aws-sdk';
const s3 = new aws.S3({
  signatureVersion: 'v4',
  region: 'eu-central-1',
});
```

We will initialize the `s3` object that we will use to upload images in the next step. It is required to pass a `region` property as a property in which we created the S3 bucket. We set the `signatureVersion` property to version `'v4'` as this is recommended.

> **Note**
>
> You can find details about the signature process of AWS requests at `https://docs.aws.amazon.com/general/latest/gr/signature-version-4.html`.

8. Inside the `resolvers.js` file, we need to add one `Upload` resolver, as follows:

```
Upload: GraphQLUpload
```

9. Inside the `mutation` property, insert the `uploadAvatar` function, as follows:

```
async uploadAvatar(root, { file }, context) {
  const { createReadStream, filename, mimetype,
    encoding } = await file;
  const bucket = 'apollo-book';
  const params = {
      Bucket: bucket,
      Key: context.user.id + '/' + filename,
      ACL: 'public-read',
      Body: createReadStream()
  };

  const response = await s3.upload(params).promise();
```

```
    return User.update({
        avatar: response.Location
    },
    {
        where: {
            id: context.user.id
        }
    }).then(() => {
        return {
            filename: filename,
            url: response.Location
        }
    });
},
```

In the preceding code snippet, we start by specifying the function as `async` so that we can use the `await` method to resolve the file and its details. The result of the resolved `await` `file` method consists of the `stream`, `filename`, `mimetype`, and `encoding` properties.

Then, we collect the following parameters in the `params` variable, in order to upload our avatar image:

- The `Bucket` field holds the name of the bucket where we save the image. I took the name `'apollo-book'`, but you will need to enter the name that you entered during the creation of the bucket. You could have specified this directly inside of the `s3` object, but this approach is a bit more flexible since you can have multiple buckets for different file types, without the need for multiple `s3` objects.

- The `Key` property is the path and name under which the file is saved. Notice that we store the file under a new folder, which is just the user **identifier (ID)** taken from the `context` variable. In a future application, you can introduce some kind of hash for every file. That would be good since the filename should not include characters that are not allowed. Furthermore, the files cannot be guessed programmatically when using a hash.

- The `ACL` field sets the permission for who can access the file. Since uploaded images on a social network are publicly viewable by anyone on the internet, we set the property to `'public-read'`.

- The `Body` field receives the `stream` variable, which we initially got by resolving the file. The `stream` variable is nothing more than the image itself as a stream, which we can directly upload into the bucket.

The `params` variable is given to the `s3.upload` function, which saves the file to our bucket. We directly chain the `promise` function onto the `upload` method. In the preceding code snippet, we use the `await` statement to resolve the promise returned by the `upload` function. Therefore, we specified the function as `async`. The `response` object of the AWS S3 upload includes the public URL under which the image is accessible to everyone.

The last step is to set the new avatar picture on the user in our database. We execute the `User.update` model function from Sequelize by setting the new URL from `response.Location`, which S3 gave us after we resolved the promise.

An example link to an S3 image is provided here:

```
https://apollo-book.s3.eu-central-1.amazonaws.com/1/test.png
```

As you can see, the URL is prefixed with the name of the bucket and then the region. The suffix is, of course, the folder, which is the user ID and the filename. The preceding URL will differ from the one that your backend generates because your bucket name and region will vary.

After updating the user, we can return the AWS response to update the **user interface** (**UI**) accordingly, without refreshing the browser window.

In the previous section, we generated access tokens in order to authorize our backend at AWS. By default, the AWS **software development kit** (**SDK**) expects both tokens to be available in our environment variables. As we did before with `JWT_SECRET`, we will set the tokens, as follows:

```
export AWS_ACCESS_KEY_ID=YOUR_AWS_KEY_ID
export AWS_SECRET_ACCESS_KEY=YOUR_AWS_SECRET_KEY
```

Insert your AWS tokens into the preceding code. The AWS SDK will detect both environment variables automatically. We do not need to read or configure them anywhere in our code.

We will now continue and implement all of the image upload features in the frontend.

React image cropping and uploading

In social networks such as Facebook, there are multiple locations where you can select and upload files. You can send images in chats, attach them to posts, create galleries in your profile, and much more. For now, we will only look at how to change our user's avatar image. This is a great example for easily showing all of the techniques.

The result that we are targeting looks like this:

Figure 7.8 – Cropping dialog

The user can select a file, crop it directly in the modal, and save it to AWS with the preceding dialog.

I am not a big fan of using too many npm packages, as this often makes your application unnecessarily big. As of the time of writing this book, we cannot write custom React components for everything, such as displaying a dialog or cropping, no matter how easy it might be.

To get the image upload working, we will install two new packages. To do this, you can follow these instructions:

1. Install the packages with npm, as follows:

    ```
    npm install --save react-modal react-dropzone react-
    cropper
    ```

    The react-modal package offers various dialog options that you can use in many different situations. The react-cropper package is a wrapper package around Cropper.js. The react-dropzone package provides an easy implementation for file drop functionality.

2. When using the `react-cropper` package, we can rely on its included **Cascading Style Sheets** (**CSS**) package. In your main `App.js` file, import it straight from the package itself, as follows:

```
import 'cropperjs/dist/cropper.css';
```

Webpack takes care of bundling all assets, as we are already doing with our custom CSS. The rest of the required CSS is available on the official GitHub repository of this book.

3. The next package that we will install is an extension for Apollo Client, which will enable us to upload files, as follows:

```
npm install --save apollo-upload-client
```

4. To get the `apollo-upload-client` package running, we have to edit the `index.js` file from the `apollo` folder where we initialize Apollo Client and all of its links. Import the `createUploadLink` function at the top of the `index.js` file, as follows:

```
import { createUploadLink } from 'apollo-upload-client';
```

5. You must replace the old `HttpLink` instance at the bottom of the link array with the new upload link. Instead of having a new `HttpLink`, we will now pass the `createUploadLink` function, but with the same parameters. When executing it, a regular link is returned. The link should look like this:

```
createUploadLink({
  uri: 'http://localhost:8000/graphql',
  credentials: 'same-origin',
}),
```

It is important to note that when we make use of the new upload link and send a file with a GraphQL request, we do not send the standard `application/json` `Content-Type` request, but instead send a multipart `FormData` request. This allows us to upload files with GraphQL. Standard **JavaScript Object Notation** (**JSON**) HTTP bodies, as we use with our GraphQL requests, cannot hold any `file` objects.

> **Note**
>
> Alternatively, it is possible to send a `base64` string instead of a `file` object when transferring images. This procedure would save you from the work that we are doing right now, as sending and receiving strings is no problem with GraphQL. You have to convert the `base64` string to a file if you want to save it in AWS S3. This approach only works for images, however, and web applications should be able to accept any file type.

6. Now that the packages are prepared, we can start to implement our uploadAvatar mutation component for the client. Create a new file called uploadAvatar.js in the mutations folder.

7. At the top of the file, import all dependencies and parse all GraphQL requests with gql in the conventional way, as follows:

```
import { gql, useMutation } from '@apollo/client';


const UPLOAD_AVATAR = gql'
  mutation uploadAvatar($file: Upload!) {
    uploadAvatar(file : $file) {
      filename
      url
    }
  }
';


export const getUploadAvatarConfig = () => ({
  update(cache, { data: { uploadAvatar } }) {
    console.log(uploadAvatar);
    if(uploadAvatar && uploadAvatar.url) {
      cache.modify({
        fields: {
          currentUser(user, { readField }) {
            cache.modify({
              id: user,
              fields: {
                avatar() {
                  return uploadAvatar.url;
                }
              }
            })
          }
        }
      });
    }
  }
});
```

```
  });
```

```
  export const useUploadAvatarMutation = () =>
  useMutation(UPLOAD_AVATAR, getUploadAvatarConfig());
```

As you can see, we have just exported the new mutation by wrapping the GraphQL query with the `useMutation` Hook. Also, we added an `update` function that will update the cache by first getting the reference for the current user and afterward updating this one user by reference to the new avatar URL.

8. Lastly, we need to add the `id` property to the `userAttributes` fragment. Otherwise, the update of the avatar URL on the user reference would only be reflected on the top bar and not with all the posts. The code is illustrated in the following snippet:

```
  import { gql } from '@apollo/client';

  export const USER_ATTRIBUTES = gql'
    fragment userAttributes on User {
      id
      username
      avatar
    }
  ';
```

The preparation is now complete. We have installed all of the required packages, configured them, and implemented the new mutation component. We can begin to program the user-facing dialog to change the avatar image.

For the purposes of this book, we are not relying on separate pages or anything like that. Instead, we are giving the user the opportunity to change their avatar when they click on their image in the top bar. To do so, we are going to listen for the click event on the avatar, opening up a dialog that includes a file dropzone and a button to submit the new image.

Execute the following steps to get this logic running:

1. It is always good to make your components as reusable as possible, so create an `avatarModal.js` file inside of the `components` folder.

2. As always, you will have to import the new `react-modal`, `react-cropper`, and `react-dropzone` packages first and then the mutation, as follows:

```
  import React, { useState, useRef } from 'react';
  import Modal from 'react-modal';
```

```
import Cropper from 'react-cropper';
import { useDropzone } from 'react-dropzone';
import { useUploadAvatarMutation } from '../apollo/
mutations/uploadAvatar';

Modal.setAppElement('#root');

const modalStyle = {
  content: {
    width: '400px',
    height: '450px',
    top: '50%',
    left: '50%',
    right: 'auto',
    bottom: 'auto',
    marginRight: '-50%',
    transform: 'translate(-50%, -50%)'
  }
};
```

As you can see in the preceding code snippet, we tell the modal package at which point in the browser's **Document Object Model** (**DOM**) we want to render the dialog, using the setAppElement method. For our use case, it is okay to take the root DOMNode, as this is the starting point of our application. The modal is instantiated in this DOMNode.

The modal component accepts a special style parameter for the different parts of the dropzone. We can style all parts of the modal by specifying the modalStyle object with the correct properties.

3. The react-cropper package gives the user the opportunity to crop the image. The result is not a file or blob object, but a dataURI object, formatted as base64. Generally, this is not a problem, but our GraphQL API expects that we send a real file, not just a string, as we explained previously. Consequently, we have to convert the dataURI object to a blob that we can send with our GraphQL request. Add the following function to take care of the conversion:

```
function dataURItoBlob(dataURI) {
  var byteString = atob(dataURI.split(',')[1]);
  var mimeString =
```

```
        dataURI.split(',')[0].split(':')[1].split(';')[0];
      var ia = new Uint8Array(byteString.length);

      for (var i = 0; i < byteString.length; i++) {
        ia[i] = byteString.charCodeAt(i);
      }


      const file = new Blob([ia], {type:mimeString});
      return file;
    }
```

Let's not get too deep into the logic behind the preceding function. The only thing that you need to know is that it converts all readable **American Standard Code for Information Interchange (ASCII)** characters into 8-bit binary data, and at the end, it returns a `blob` object to the calling function. It converts data URIs to blobs.

4. The new component that we are implementing at the moment is called `AvatarUpload`. It receives the `isOpen` property, which sets the modal to visible or invisible. By default, the modal is invisible. Furthermore, when the modal is shown, the dropzone is rendered inside. First, set up the component itself and the required variables, as follows:

```
const AvatarModal = ({ isOpen, showModal }) => {
  const [file, setFile] = useState(null);
  const [result, setResult] = useState(null);
  const [uploadAvatar] = useUploadAvatarMutation();
  const cropperRef = useRef(null);
}
```

We require the `file` and `result` state variables to manage the original file selected and the cropped image. Furthermore, we set up the mutation and a reference using the `useRef` Hook, which is required for the `cropper` library.

5. Next, we need to set up all the component functions that we will use to handle different events and callbacks. Add the following functions to the component:

```
const saveAvatar = () => {
  const resultFile = dataURItoBlob(result);
  resultFile.name = file.filename;
  uploadAvatar({variables: { file: resultFile
    }}).then(() => {
```

```
      showModal();
    });
  };

  const changeImage = () => {
    setFile(null);
  };

  const onDrop = (acceptedFiles) => {
    const reader = new FileReader();
    reader.onload = () => {
      setFile({
        src: reader.result,
        filename: acceptedFiles[0].name,
        filetype: acceptedFiles[0].type,
        result: reader.result,
        error: null,
      });
    };
    reader.readAsDataURL(acceptedFiles[0]);
  };
  const {getRootProps, getInputProps, isDragActive} =
  useDropzone({onDrop});

  const onCrop = () => {
    const imageElement = cropperRef?.current;
    const cropper = imageElement?.cropper;
    setResult(cropper.getCroppedCanvas().toDataURL());
  };
```

The saveAvatar function is the main function that will translate the base64 string into a blob. The onDrop function is called when the user drops or selects an image. At this moment, we use FileReader to read the file and give us the base64 string that we save in the file state variable as an object. The useDropZone Hook gives us all the properties that we can use to set up the actual dropzone.

The changeImage function will cancel the current crop process and allow us to upload a new file again.

The onCrop function is called every time the cropped selection is changed by the user. At this moment, we save the new cropped image as a base64 string to the result state variable to have a clear separation between the original file variable and the result variable.

6. The Modal component takes an onRequestClose method, which executes the showModal function when the user tries to close the modal by clicking outside of it, for example. We receive the showModal function from the parent component, which we are going to cover in the next step. The modal also receives the default style property and a label.

The Cropper component needs to receive a function in the crop property that is called on every change. Also, the Cropper component receives the src property from the file state variable, as illustrated in the following code snippet:

```
return (
  <Modal
    isOpen={isOpen}
    onRequestClose={showModal}
    contentLabel="Change avatar"
    style={modalStyle}
  >
    {!file &&
      (<div className="drop" {...getRootProps()}>
        <input {...getInputProps()} />
        {isDragActive ? <p>Drop the files here ...</p>
        : <p>Drag 'n' drop some files here, or click
        to select files</p>}
      </div>)
    }
    {file && <Cropper ref={cropperRef}
    src={file.src} style={{ height: 400, width:
    "100%" }} initialAspectRatio={16 / 9}
    guides={false} crop={onCrop}/>}
    {file && (
```

```
            <button className="cancelUpload"
                onClick={changeImage}>Change image</button>
        )}
        <button className="uploadAvatar"
            onClick={saveAvatar}>Save</button>
    </Modal>
    )
```

The `return` statement, as you can see, only includes the modal as a wrapper and a cropper. At the end, we have a button calling `saveAvatar` to execute the mutation, and with it send the cropped image or `changeImage`, which cancels the cropping for the current image.

7. Don't forget to add the `export` statement to the end of the file, as follows:

```
export default AvatarModal
```

8. Now, switch over to the `user.js` file in the `bar` folder, where all of the other application bar-related files are stored. Import the new `AvatarModal` component, as follows:

```
import AvatarModal from '../avatarModal';
```

9. The `UserBar` component is the parent of `AvatarUploadModal`. Open the `user.js` file from the `bar` folder. That is why we handle the `isOpen` state variable of the dialog in the `UserBar` component. We introduce an `isOpen` state variable and catch the `onClick` event on the avatar of the user. Copy the following code into the `UserBar` component:

```
const [isOpen, setIsOpen] = useState(false);

const showModal = () => {
    setIsOpen(!isOpen);
}
```

10. Replace the `return` statement with the following code:

```
return (
    <div className="user">
        <img src={user.avatar} onClick={() => showModal()} />
        <AvatarModal isOpen={isOpen}
```

```
            showModal={showModal}/>
        <span>{user.username}</span>
    </div>
 );
```

The modal component directly receives the `isOpen` property, as we explained earlier. The `showModal` method is executed when the avatar image is clicked. This function updates the property of the `AvatarModal` component, and either shows or hides the modal.

Start the server and client with the matching `npm run` commands. Reload your browser and try out the new feature. When an image is selected, the cropping tool is displayed. You can drag and resize the image area that should be uploaded. You can see an example of this in the following screenshot:

Figure 7.9 – Cropping in progress

Hitting **Save** uploads the image under the `user` folder in the S3 bucket. Thanks to the mutation that we wrote, the avatar image in the top bar is updated with the new URL to the S3 bucket location of the image.

The great thing that we have accomplished is that we send the images to our server. Our server transfers all of the images to S3. AWS responds with the public URL, which is then placed directly into the avatar field in the browser. The way that we query the avatar image from the backend, using our GraphQL API, does not change. We return the URL to the S3 file, and everything works.

Summary

In this chapter, we started by creating an AWS account and an S3 bucket for uploading static images from our backend. Modern social networks consist of many images, videos, and other types of files. We introduced Apollo Client, which allows us to upload any type of file. In this chapter, we managed to upload an image to our server, and we covered how to crop images and save them through a server in AWS S3. Your application should now be able to serve your users with images at any time.

The next chapter will cover the basics of client-side routing, with the use of React Router.

8
Routing in React

Currently, we have one screen and one path that our users can visit. When users visit Graphbook, they can log in and see their news feed and chats. Another requirement for a social network is that users have their own profile pages. We will implement this feature in this chapter.

We will introduce client-side routing for our React application.

This chapter will cover the following topics:

- Setting up React Router
- Advanced routing with React Router

Technical requirements

The source code for this chapter is available in the following GitHub repository:

https://github.com/PacktPublishing/Full-Stack-Web-Development-with-GraphQL-and-React-Second-Edition/tree/main/Chapter08

Setting up React Router

Routing is essential to most web applications. You cannot cover all of the features of your application in just one page. It would be overloaded, and your user would find it difficult to understand. Sharing links to pictures, profiles, or posts is also very important for a social network such as Graphbook. One advantageous feature, for example, is being able to send links to specific profiles. This requires each profile to have its own **Uniform Resource Locator** (**URL**) and page. Otherwise, it will not be possible to share a direct link to a single item of your application. It is also crucial to split the content into different pages, due to **search engine optimization** (**SEO**).

At the moment, we render our complete application to **HyperText Markup Language** (**HTML**) in the browser, based on the authentication status. Only the server implements a simple routing functionality. Carrying out client-side routing can save a lot of work and time for the user if the router merely swaps out the correct parts in React, instead of reloading the page completely when following a link. It is vital that the application makes use of the HTML5 history implementation so that it handles the history of the browser. Importantly, this should also work for navigation in different directions. We should be able to go forward and backward with the arrow navigation buttons in the browser, without reloading the application. No unnecessary page reloads should happen with this solution.

Common frameworks that you may know about, such as Angular, Ember, and Ruby on Rails, use static routing. That is also the case for Express.js, which we covered in *Chapter 2, Setting Up GraphQL with Express.js*, of this book. **Static routing** means that you configure your routing flow and the components to render upfront. Your application then processes the routing table in a separate step, renders the required components, and presents the results to the user.

With the release of version 4 and also the current version 5 of React Router, which we are going to use, **dynamic routing** was introduced. The unique thing about it is that the routing takes place while the rendering of your application is running. It doesn't require the application to first process a configuration in order to show the correct components. This approach fits with React's workflow well. The routing happens directly in your application, not in a preprocessed configuration.

Installing React Router

In the past, there were a lot of React routers, with various implementations and features. As we mentioned previously, we are going to install and configure version 5 for this book. If you search for other tutorials on this topic, make sure that you follow the instructions for this version. Otherwise, you might miss some of the changes that React Router has gone through.

To install React Router, simply run npm again, as follows:

```
npm install --save react-router-dom
```

From the package name, you might assume that this is not the main package for React. The reason for this is that React Router is a multi-package library. That comes in handy when using the same tool for multiple platforms. The core package is called react-router.

There are two further packages. The first one is the react-router-dom package, which we installed in the preceding code snippet, and the second one is the react-router-native package. If at some point, you plan to build a React Native app, you can use the same routing instead of using the browser's **Document Object Model (DOM)** for a real mobile app.

The first step that we will take introduces a simple router to get our current application working, including different paths for all of the screens. The routes that we are going to add are detailed here:

- Our posts feed, chats, and the top bar, including the search box, should be accessible under the /app route of our application. The path is self-explanatory, but you could also use the / root as the main path.

- The login and signup forms should have a separate path, which will be accessible under the / root path.

- As we do not have any further screens, we also have to handle a situation in which none of the preceding routes match. In that case, we could display a so-called 404 page, but instead, we are going to redirect to the root path directly.

There is one thing that we have to prepare before continuing. For development, we are using the webpack development server, as this is what we configured in *Chapter 1, Preparing Your Development Environment*. To get the routing working out of the box, we will add two parameters to the webpack.client.config.js file. The devServer field should look like this:

```
devServer: {
  port: 3000,
  open: true,
  historyApiFallback: true,
},
```

The `historyApiFallback` field tells `devServer` to serve the `index.html` file, not only for the root path, `http://localhost:3000/`, but also when it would typically receive a 404 error (such as for paths like `http://localhost:3000/app`). This happens when the path does not match a file or folder that is normal when implementing routing.

The `output` field at the top of the `config` file must have a `publicPath` property, as follows:

```
output: {
  path: path.join(__dirname, buildDirectory),
  filename: 'bundle.js',
  publicPath: '/',
},
```

The `publicPath` property tells webpack to prefix the bundle URL to an absolute path, instead of a relative path. When this property is not included, the browser cannot download the bundle when visiting the sub-directories of our application, as we are implementing client-side routing. Let's begin with the first path and bind the central part of our application, including the news feed, to the `/app` path.

Implementing your first route

Before implementing the routing, we will clean up the `App.js` file. To do this, follow these steps:

1. Create a `Main.js` file next to the `App.js` file in the `client` folder. Insert the following code:

   ```
   import React from 'react';
   import Feed from './Feed';
   import Chats from './Chats';
   import Bar from './components/bar';

   export const Main = ({ changeLoginState }) => {
     return (
       <>
         <Bar changeLoginState={changeLoginState} />
         <Feed />
         <Chats />
       </>
   ```

```
    );
  }
```

```
  export default Main;
```

As you might have noticed, the preceding code is pretty much the same as the logged-in condition inside the App.js file. The only change is that the changeLoginState function is taken from the properties and is not directly a method of the component itself. That is because we split this part out of the App.js file and put it in a separate file. This improves reusability for other components that we are going to implement.

2. Now, open and replace the return statement of the App component to reflect those changes, as follows:

```
  return (
    <div className="container">
      <Helmet>
        <title>Graphbook - Feed</title>
        <meta name="description" content="Newsfeed of
          all your friends on Graphbook" />
      </Helmet>
      <Router loggedIn={loggedIn}
        changeLoginState={handleLogin}/>
    </div>
  )
```

If you compare the preceding method with the old one, you can see that we have inserted a Router component, instead of directly rendering either the posts feed or the login form. The original components of the App.js file are now in the previously created Main.js file. Here, we pass the loggedIn property and the changeLoginState function to the Router component. Remove the dependencies at the top, such as the Chats and Feed components, because we won't use them any more thanks to the new Main component.

3. Add the following line to the dependencies of our App.js file:

```
  import Router from './router';
```

To get the routing working, we have to implement our custom `Router` component first. Generally, it is easy to get the routing running with React Router, and you are not required to separate the routing functionality into a separate file, but that makes it more readable.

4. To do this, create a new `router.js` file in the `client` folder, next to the `App.js` file, with the following content:

```
import React from 'react';
import { BrowserRouter as Router, Route, Redirect, Switch
} from 'react-router-dom';
import LoginRegisterForm from './components/
loginregister';
import Main from './Main';

export const routing = ({ changeLoginState, loggedIn })
=> {
  return (
    <Router>
      <Switch>
        <Route path="/app" component={() => <Main
          changeLoginState= {changeLoginState}/>}/>
      </Switch>
    </Router>
  )
}

export default routing;
```

At the top, we import all of the dependencies. They include the new `Main` component and the `react-router` package. Here is a quick explanation of all of the components that we are importing from the React Router package:

*   `BrowserRouter` (or `Router`, for short, as we called it here) is the component that keeps the URL in the address bar in sync with the **user interface** (**UI**); it handles all of the routing logic.

*   The `Switch` component forces the first matching `Route` or `Redirect` component to be rendered. We need it to stop re-rendering the UI if the user is already in the location to which a redirect is trying to navigate. I generally recommend that you use the `Switch` component, as it catches unforeseeable routing errors.

- `Route` is the component that tries to match the given path to the URL of the browser. If this is the case, the `component` property is rendered. You can see in the preceding code snippet that we are not setting the `Main` component directly as a parameter; instead, we return it from a stateless function. That is required because the `component` property of a `Route` component only accepts functions and not a component object. This solution allows us to pass the `changeLoginState` function to the `Main` component.

- `Redirect` navigates the browser to a given location. The component receives a property called `to`, filled by a path starting with `/`. We are going to use this component in the next section.

The problem with the preceding code is that we are only listening for one route, which is `/app`. If you are not logged in, there will be many errors that are not covered. The best thing to do would be to redirect the user to the root path, where they can log in.

Secured routes

Secured routes represent a way to specify paths that are only accessible if the user is authenticated or has the correct authorization.

The recommended solution to implement secure routes in React Router is to write a small, stateless function that conditionally renders either a `Redirect` component or the component specified on the route that requires an authenticated user. We extract the `component` property of the router into the `Component` variable, which is a renderable React object:

1. Insert the following code into the `router.js` file:

```
const PrivateRoute = ({ component: Component, ...rest })
=> (
  <Route {...rest} render={ (props) => (
    rest.loggedIn === true
      ? <Component {...props} />
      : <Redirect to={{
          pathname: '/',
      }} />
  )} />
)
```

We call the `PrivateRoute` stateless function. It returns a standard `Route` component, which receives all of the properties initially given to the `PrivateRoute` function. To pass all properties, we use a destructuring assignment with the `...rest` syntax. Using the syntax inside of curly braces on a React component passes all fields of the `rest` object as properties to the component. The `Route` component is only rendered if the given path is matched.

Furthermore, the rendered component is dependent on the user's `loggedIn` state variable, which we have to pass. If the user is logged in, we render the `Component` variable without any problems. Otherwise, we redirect the user to the root path of our application using the `Redirect` component.

2. Use the new `PrivateRoute` component in the `return` statement of the `Router` component and replace the old `Route` component, as follows:

```
<PrivateRoute path="/app" component={() =>
<Main changeLoginState={changeLoginState} />}
loggedIn={loggedIn}/>
```

Notice that we pass the `loggedIn` property by taking the value from the properties of the `Router` component itself. It initially receives the `loggedIn` property from the App component that we edited previously. The great thing is that the `loggedIn` variable can be updated from the parent App component at any time. That means that the `Redirect` component is rendered, and the user is automatically navigated to the login form if the user logs out, for example. We do not have to write separate logic to implement this functionality.

However, we have now created a new problem. We redirect from `/app` to `/` if the user is not logged in, but we do not have any routes set up for the initial `'/'` path. It makes sense for this path to either show the login form or to redirect the user to `/app` if the user is logged in. The pattern for the new component is the same as the preceding code for the `PrivateRoute` component but in the opposite direction.

3. Add the new `LoginRoute` component to the `router.js` file, as follows:

```
const LoginRoute = ({ component: Component, ...rest }) =>
(
  <Route {...rest} render={(props) => (
    rest.loggedIn === false
      ? <Component {...props} />
      : <Redirect to={{
          pathname: '/app',
        }} />
```

```
    )} />
  )
```

The preceding condition is inverted to render the original component. If the user is not logged in, the login form is rendered. Otherwise, they will be redirected to the posts feed.

4. Add the new path to the router, as follows:

```
<LoginRoute exact path="/" component={() =>
<LoginRegisterForm changeLoginState={changeLoginState}/>}
loggedIn={loggedIn}/>
```

The code looks the same as that of the `PrivateRoute` component, except that we now have a new property, called `exact`. If we pass this property to a route, the browser's location has to match 100%. The following table shows a quick example, taken from the official React Router documentation:

Router path	Browser path	Exact	Matches
/one	/one/two	True	No
/one	/one/two	False	Yes

For the root path, we set `exact` to `true` because otherwise, the path matches with any browser's location where / is included, as you can see in the preceding table.

> **Note**
> There are many more configuration options that React Router offers, such as enforcing trailing slashes, case sensitivity, and much more. You can find all of the options and examples in the official documentation at `https://v5.reactrouter.com/web/api/`.

Catch-all routes in React Router

Currently, we have two paths set up, which are `/app` and `/`. If a user visits a non-existent path, such as `/test`, they will see an empty screen. The solution is to implement a route that matches any path. For simplicity, we redirect the user to the root of our application, but you could easily replace the redirection with a typical 404 page.

Add the following code to the `router.js` file:

```
const NotFound = () => {
  return (
    <Redirect to="/"/>
  );
}
```

The `NotFound` component is minimal. It just redirects the user to the root path. Add the next `Route` component to the `Switch` component in the `Router` component. Ensure that it is the last one on the list. The code is shown here:

```
<Route component={NotFound} />
```

As you can see, we are rendering a simple `Route` component in the preceding code. What makes the route special is that we are not passing a `path` property with it. By default, the `path` property is completely ignored, and the component is rendered every time, except if there is a match with a previous component. That is why we added the route to the bottom of the `Router` component. When no route matches, we redirect the user to the login screen in the root path, or, if the user is already logged in, we redirect them to a different screen using the routing logic of the root path. Our `LoginRoute` component handles this last case.

You can test all changes when starting the frontend with `npm run client` and the backend with `npm run server`. We have now moved the current state of our application from a standard, single-route application to an application that differentiates the login form and the news feed based on the location of the browser.

In the next section, we will have a look at how we can implement more complicated routing by adding parameterized routes and loading the data depending on those parameters.

Advanced routing with React Router

The primary goal of this chapter is to build a profile page for your users. We need a separate page to show all of the content that a single user has entered or created. The content would not fit next to the posts feed. When looking at Facebook, we can see that every user has their own address, under which we can find the profile page of a specific user. We are going to create our profile page in the same way and use the username as the custom path.

We have to implement the following features:

1. We add a new parameterized route for the user profile. The path starts with /user/ and follows a username.

2. We change the user profile page to send all GraphQL queries, including the username route parameter, inside of the variables field of the GraphQL request.

3. We edit the postsFeed query to filter all posts by the username parameter provided.

4. We implement a new GraphQL query on the backend to request a user by their username, in order to show information about the user.

5. When all of the queries are finished, we render a new user profile header component and the posts feed.

6. Finally, we enable navigation between each page without reloading the complete page, but only the changed parts.

Let's start by implementing routing for the profile page in the next section.

Parameters in routes

We have prepared most of the work required to add a new user route. Open up the router.js file again. Add the new route, as follows:

```
<PrivateRoute path="/user/:username" component={props =>
<User {...props} changeLoginState={changeLoginState}/>}
loggedIn={loggedIn}/>
```

The code contains two new elements, as follows:

- The path that we entered is /user/:username. As you can see, the username is prefixed with a colon, telling React Router to pass the value of it to the underlying component being rendered.

- The component that we rendered previously was a stateless function that returned either the LoginRegisterForm component or the Main component. Neither of these received any parameters or properties from React Router. Now, however, it is required that all properties of React Router are transferred to the child component. That includes the username parameter that we just introduced. We use the same destructuring assignment with the props object to pass all properties to the User component.

Those are all of the changes that we need to accept parameterized paths in React Router. We read out the value inside of the new user page component. Before implementing it, we import the dependency at the top of `router.js` to get the preceding route working, as follows:

```
import User from './User';
```

Create the preceding `User.js` file next to the `Main.js` file. As with the `Main` component, we are collecting all of the components that we render on this page. You should stay with this layout, as you can directly see which main parts each page consists of. The `User.js` file should look like this:

```
import React from 'react';
import UserProfile from './components/user';
import Chats from './Chats';
import Bar from './components/bar';

export const User = ({ changeLoginState, match }) => {
  return (
    <>
      <Bar changeLoginState={changeLoginState} />
      <UserProfile username={match.params.username}/>
      <Chats />
    </>
  );
}

export default User
```

We have all common components including the `Bar` and `Chat` component. If a user visits the profile of a friend, they see the common application bar at the top. They can access their chats on the right-hand side, like on Facebook. It is one of the many situations in which React and the reusability of components come in handy.

We removed the `Feed` component and replaced it with a new `UserProfile` component. Importantly, the `UserProfile` component receives the `username` property. Its value is taken from the properties of the `User` component. These properties were passed over by React Router. If you have a parameter, such as `username`, in the routing path, the value is stored in the `match.params.username` property of the child component. The `match` object generally contains all matching information of React Router.

From this point on, you can implement any custom logic that you want with this value. We will now continue with implementing the profile page.

One thing before building the user profile page is to extract the feed rendering logic to a separate component to reuse it. Create a `feedlist.js` file inside the `post` folder.

Insert the following code in the `feedlist.js` file:

1. Import the following dependencies at the top, as follows:

    ```
    import React, { useState } from 'react';
    import InfiniteScroll from 'react-infinite-scroll-
    component';
    import Post from './';
    ```

2. Then, just copy the main parts of the feed list `return` statement, as follows:

    ```
    export const FeedList = ({fetchMore, posts}) => {
      const [hasMore, setHasMore] = useState(true);
      const [page, setPage] = useState(0);

      return (
        <div className="feed">
          <InfiniteScroll
            dataLength={posts.length}
            next={() => loadMore(fetchMore)}
            hasMore={hasMore}
            loader={<div className="loader"
              key={"loader"}>Loading ...</div>}
          >
          {posts.map((post, i) =>
              <Post key={post.id} post={post} />
          )}
          </InfiniteScroll>
        </div>
      );
    }

    export default FeedList;
    ```

3. One thing that is missing now is the `loadMore` function, which we also can just copy. Just add it straight to the preceding component, as follows:

```
const loadMore = (fetchMore) => {
    fetchMore({
      variables: {
        page: page + 1,
      },
      updateQuery(previousResult, { fetchMoreResult })
      {
        if(!fetchMoreResult.postsFeed.posts.length) {
          setHasMore(false);
          return previousResult;
        }

        setPage(page + 1);

        const newData = {
          postsFeed: {
            __typename: 'PostFeed',
            posts: [
              ...previousResult.postsFeed.posts,
              ...fetchMoreResult.postsFeed.posts
            ]
          }
        };
        return newData;
      }
    });
}
```

4. Just replace the part on the `Feed.js` file's `return` statement. It should look like this:

```
return (
  <div className="container">
    <div className="postForm">
      <form onSubmit={handleSubmit}>
```

```
                    <textarea value={postContent} onChange={ (e) =>
                        setPostContent(e.target.value)}
                            placeholder="Write your custom post!"/>
                    <input type="submit" value="Submit" />
                </form>
            </div>
            <FeedList posts={posts} fetchMore={loadMore}/>
        </div>
    )
```

We can use this `FeedList` component now where we need to display a feed of posts, such as on our user profile page.

Follow these steps to build the user's profile page:

1. Create a new folder, called `user`, inside the `components` folder.
2. Create a new file, called `index.js`, inside the `user` folder.
3. Import the dependencies at the top of the file, as follows:

```
import React from 'react';
import FeedList from '../post/feedlist';
import UserHeader from './header';
import Loading from '../loading';
import Error from '../error';
import { useGetPostsQuery } from '../../apollo/queries/
getPosts';
import { useGetUserQuery } from '../../apollo/queries/
getUser';
```

The first three lines should look familiar. Two imported files, however, do not exist at the moment, but we are going to change that shortly. The first new file is `UserHeader`, which takes care of rendering the avatar image, the name, and information about the user. Logically, we request the data that we will display in this header through a new Apollo query Hook, called `getUser`.

4. Insert the code for the `UserProfile` component that we are building at the moment beneath the dependencies, as follows:

```
const UserProfile = ({ username }) => {
    const { data: user, loading: userLoading } =
        useGetUserQuery({ username });
```

```
    const { loading, error, data: posts, fetchMore } =
      useGetPostsQuery({ username });

    if (loading || userLoading) return <Loading />;
    if (error) return <Error><p>{error.message}</p>
      </Error>;

    return (
      <div className="user">
        <div className="inner">
          <UserHeader user={user.user} />
        </div>
        <div className="container">
          <FeedList posts={posts.postsFeed.posts}
            fetchMore={fetchMore}/>
        </div>
      </div>
    )
  }

  export default UserProfile;
```

The `UserProfile` component is not complex. We are running two Apollo queries simultaneously. Both have the `variables` property set. The `useGetPostsQuery` Hook receives the username, which initially came from React Router. This property is also handed over to `useGetUserQuery`.

5. Now edit and create Apollo queries, before programming the profile header component. Open the `getPosts.js` file from the `queries` folder.

6. To use the username as input to the GraphQL query, we first have to change the query string from the `GET_POSTS` variable. Change the first two lines to match the following code:

```
query postsFeed($page: Int, $limit: Int, $username:
String) {
  postsFeed(page: $page, limit: $limit, username:
    $username) {
```

7. Then, replace the last line with the following code to provide a way to pass variables to the `useQuery` Hook:

```
export const useGetPostsQuery = (variables) =>
useQuery(GET_POSTS, { pollInterval: 5000, variables: {
page: 0, limit: 10, ...variables } });
```

If the custom query component receives a `username` property, it is included in the GraphQL request. It is used to filter posts by the specific user that we are viewing.

8. Create a new `getUser.js` file in the `queries` folder to create a query Hook, which we are missing at present.

9. Import all of the dependencies and parse the new query schema with `gql`, as follows:

```
import { gql, useQuery } from '@apollo/client';
import { USER_ATTRIBUTES } from '../fragments/
userAttributes';


export const GET_USER = gql'
  query user($username: String!) {
    user(username: $username) {
      ...userAttributes
    }
  }
  ${USER_ATTRIBUTES}
';


export const useGetUserQuery = (variables) =>
useQuery(GET_USER, { variables: { ...variables }});
```

The preceding query is nearly the same as the `currentUser` query. We are going to implement the corresponding `user` query later, in our GraphQL **application programming interface** (**API**).

10. The last step is to implement the `UserProfileHeader` component. This component renders the `user` property, with all its values. It is just simple HTML markup. Copy the following code into the `header.js` file, in the `user` folder:

```
import React from 'react';


export const UserProfileHeader = ({user}) => {
```

```
      const { avatar, username } = user;

    return (
      <div className="profileHeader">
        <div className="avatar">
          <img src={avatar}/>
        </div>
        <div className="information">
          <p>{username}</p>
          <p>You can provide further information here
            and build your really personal header
            component for your users.</p>
        </div>
      </div>
    )
  }

  export default UserProfileHeader;
```

If you need help getting the **Cascading Style Sheets** (**CSS**) styling right, take a look at the official repository for this book. The preceding code only renders the user's data; you could also implement features such as a chat button, which would give the user the option to start messaging with other people. Currently, we have not implemented this feature anywhere, but it is not necessary to explain the principles of React and GraphQL.

We have finished the new frontend components, but the `UserProfile` component is still not working. The queries that we are using here either do not accept the `username` parameter or have not yet been implemented.

The next section will cover which parts of the backend have to be adjusted.

Querying the user profile

With the new profile page, we have to update our backend accordingly. Let's take a look at what needs to be done, as follows:

- We have to add the `username` parameter to the schema of the `postsFeed` query and adjust the resolver function.

- We have to create a schema and a resolver function for the new `UserQuery` component.

We will begin with the `postsFeed` query:

1. Edit the `postsFeed` query in the `RootQuery` type of the `schema.js` file to match the following code:

```
postsFeed(page: Int, limit: Int, username: String):
PostFeed @auth
```

Here, I have added `username` as an optional parameter.

2. Now, head over to the `resolvers.js` file and take a look at the corresponding `resolver` function. Replace the signature of the function to extract the username from the variables, as follows:

```
postsFeed(root, { page, limit, username }, context) {
```

3. To make use of the new parameter, add the following lines of code above the `return` statement:

```
if (username) {
  query.include = [{model: User}];
  query.where = { '$User.username$': username };
}
```

We have already covered the basic Sequelize API and how to query associated models by using the `include` parameter in *Chapter 3, Connecting to the Database*. An important point is how we filter posts associated with a user by their username. We'll do this in the following steps:

1. In the preceding code, we fill the `include` field of the `query` object with the Sequelize model that we want to join. This allows us to filter the associated `User` model in the next step.

2. Then, we create a normal `where` object in which we write the filter condition. If you want to filter the posts by an associated table of users, you can wrap the model and field names that you want to filter by with dollar signs. In our case, we wrap `User.username` with dollar signs, which tells Sequelize to query the `User` model's table and filter by the value of the `username` column.

No adjustments are required for the pagination part. The GraphQL query is now ready. The great thing about the small changes that we have made is that we have just one API function that accepts several parameters, either to display posts on a single user profile or to display a list of posts, such as a news feed.

Let's move on and implement the new `user` query:

1. Add the following line to the `RootQuery` type in your GraphQL schema:

   ```
   user(username: String!): User @auth
   ```

   This query only accepts a `username` parameter, but this time it is a required parameter in the new query. Otherwise, the query would make no sense since we only use it when visiting a user's profile through their username.

2. In the `resolvers.js` file, implement the resolver function using Sequelize, as follows:

   ```
   user(root, { username }, context) {
     return User.findOne({
       where: {
         username: username
       }
     });
   },
   ```

   In the preceding code snippet, we use the `findOne` method of the `User` model by using Sequelize and search for exactly one user with the username that we provided in the parameter.

Now that the backend code and the user page are ready, we have to allow the user to navigate to this new page. The next section will cover user navigation using React Router.

Programmatic navigation in React Router

We created a new site with the user profile, but now we have to offer the user a link to get there. The transition between the news feed and the login and registration forms is automated by React Router, but not the transition from the news feed to a profile page. The user decides whether they want to view the profile of the user. React Router has multiple ways to handle navigation. We are going to extend the news feed to handle clicks on the username or the avatar image, in order to navigate to the user's profile page. Open the `header.js` file in the `post` components folder. Import the `Link` component provided by React Router, as follows:

```
import { Link } from 'react-router-dom';
```

The `Link` component is a tiny wrapper around a regular HTML a tag. Apparently, in standard web applications or websites, there is no complex logic behind hyperlinks; you click on them, and a new page is loaded from scratch. With React Router or most **single-page application (SPA) JavaScript (JS)** frameworks, you can add more logic behind hyperlinks. Importantly, instead of completely reloading the pages when navigating between different routes, this now gets handled by React Router. There won't be complete page reloads when navigating; instead, only the required parts are exchanged, and the GraphQL queries are run. This method saves the user expensive bandwidth because it means that we can avoid downloading all of the HTML, CSS, and image files again.

To test this, wrap the username and the avatar image in the `Link` component, as follows:

```
<Link to={'/user/'+post.user.username}>
  <img src={post.user.avatar} />
  <div>
    <h2>{post.user.username}</h2>
  </div>
</Link>
```

In the rendered HTML, the `img` and `div` tags are surrounded by a common a tag but are handled inside React Router. The `Link` component receives a `to` property, which is the destination of the navigation. You have to copy one new CSS rule because the `Link` component has changed the markup. The code is illustrated in the following snippet:

```
.post .header a > * {
  display: inline-block;
  vertical-align: middle;
}
```

If you test the changes now, clicking on the username or avatar image, you should notice that the content of the page dynamically changes but does not entirely reload. A further task would be to copy this approach to the user search list in the application bar and the chats. Currently, the users are displayed, but there is no option to visit their profile pages by clicking on them.

Now, let's take a look at another way to navigate with React Router. If the user has reached a profile page, we want them to navigate back by clicking on a button in the application bar. First of all, we will create a new `home.js` file in the `bar` folder, and we will enter the following code:

```
import React from 'react';
import { withRouter } from 'react-router';
```

```
const Home = ({ history }) => {
  const goHome = () => {
    history.push('/app');
  }

  return (
    <button className="goHome" onClick={goHome}>Home
    </button>
  );
}


export default withRouter(Home);
```

We are using multiple React Router techniques here. We export the Home component through a **higher-order component (HOC)**. The withRouter HOC gives the Home component access to the history object of React Router. That is great because it means that we do not need to pass this object from the top of our React tree down to the Home component.

Furthermore, we use the history object to navigate the user to the news feed. In the render method, we return a button that, when clicked, runs the history.push function. This function adds the new path to the history of the browser and navigates the user to the '/app' main page. The good thing is that it works in the same way as the Link component and does not reload the entire website.

There are a few things to do in order to get the button working, as follows:

1. Import the component into the index.js file of the bar folder, as follows:

    ```
    import Home from './home';
    ```

2. Then, replace the buttons div tag with the following lines of code:

    ```
    <div className="buttons">
      <Home/>
      <Logout changeLoginState={changeLoginState}/>
    </div>
    ```

3. Wrap the two buttons in a separate div tag so that it is easier to align them correctly. You can replace the old CSS for the **Logout** button and add the following:

    ```
    .topbar .buttons {
      position: absolute;
    ```

```
    right: 5px;
    top: 5px;
    height: calc(100% - 10px);
}

.topbar .buttons > * {
    height: 100%;
    margin-right: 5px;
    border: none;
    border-radius: 5px;
}
```

Now that we have everything together, the user can visit the profile page and navigate back again. Our final result looks like this:

Figure 8.1 – User profile

We have a big profile header for the user and their posts at the bottom of the window. At the top, you can see the top bar with the currently logged-in user.

Remembering the redirect location

When a visitor comes to your page, they have probably followed a link that was posted elsewhere. This link is likely to be a direct address for a user, a post, or anything else that you offer direct access to. For those that are not logged in, we configured the application to redirect that person to the login or signup forms. This behavior makes sense. However, once that person has either logged in or signed up with a new account, they are then navigated to the news feed. A better way of doing this would be to remember the initial destination that the person wanted to visit. To do this, we will make a few changes to the router. Open the `router.js` file. With all of the routing components provided by React Router, we always get access to the properties inside of them. We will make use of this and save the last location that we were redirected from.

In the `PrivateRoute` component, swap out the `Redirect` component with the following code:

```
<Redirect to={{
    pathname: '/',
    state: { from: props.location }
}} />
```

Here, we have added the `state` field. The value that it receives comes from the parent `Route` component, which holds the last matched path in the `props.location` field generated by React Router. The path can be a user's profile page or the news feed since both rely on the `PrivateRoute` component where authentication is required. When the preceding redirect is triggered, you receive the `from` field inside the router's state.

We want to use this variable when the user is logging in. Replace the `Redirect` component in the `LoginRoute` component with the following lines of code:

```
<Redirect to={{
    pathname: (typeof props.location.state !== typeof
        undefined) ?
    props.location.state.from.pathname : '/app',
}} />
```

Here, I have introduced a small condition for the `pathname` parameter. If the `location.state` property is defined, we can rely on the `from` field. Previously, we stored the redirect path in the `PrivateRoute` component. If the `location.state` property does not exist, the user was not visiting a direct hyperlink but just wanted to log in normally. They will be navigated to the news feed with the `/app` path.

Your application should now be able to handle all routing scenarios, and this should allow your users to view your site comfortably.

Summary

In this chapter, we transitioned from our one-screen application to a multi-page setup. React Router, our main library for routing purposes now has three paths, under which we display different parts of Graphbook. Furthermore, we now have a catch-all route, in which we can redirect the user to a valid page.

In the next chapter, we will continue with this progression by implementing server-side rendering, which needs many adjustments on both the frontend and the backend.

9

Implementing Server-Side Rendering

With our progress from the last chapter, we are now serving multiple pages under different paths with our **React** application. Currently, all of the routing happens directly on the client. In this chapter, we will look at the advantages and disadvantages of **server-side rendering** (**SSR**). By the end of the chapter, you will have configured **Graphbook** to serve all pages as pre-rendered HTML from the server instead of the client.

This chapter covers the following topics:

- Introducing SSR
- Setting up SSR in **Express.js** to render React on the server
- Enabling **JSON Web Token** (**JWT**) authentication in connection with SSR
- Running all of our **GraphQL** queries in the React tree

Technical requirements

The source code for this chapter is available in the following **GitHub** repository:

`https://github.com/PacktPublishing/Full-Stack-Web-Development-with-GraphQL-and-React-Second-Edition/tree/main/Chapter09`

Introducing SSR

First, you have to understand the differences between using a server-side-rendered and a client-side-rendered application. There are numerous things to bear in mind when transforming a pure client-side-rendered application to support SSR. In our application, the current user flow begins with the client requesting a standard `index.html` file. This file includes only a small number of things, such as a small `body` object with one `div` element, a `head` tag with some very basic `meta` tags, and a vital `script` tag that downloads the bundled **JavaScript** file created by **webpack**. The server merely serves the `index.html` and the `bundle.js` files. Then, the client's browser begins processing the React markup that we wrote. When React has finished evaluating the code, we see the HTML of the application that we wanted to see. All CSS files or images are also downloaded from our server, but only when React has inserted the HTML into the browser's **Document Object Model** (**DOM**). During the rendering by React, the **Apollo** components are executed and all of the queries are sent. These queries are, of course, handled by our backend and database.

In comparison with SSR, the client-side approach is straightforward. Before the development of **Angular**, **Ember**, **React**, and other JavaScript frameworks, the conventional approach was to have a backend that implemented all of the business logic and also a high number of templates or functions that returned valid HTML. The backend queried the database, processed the data, and inserted the data into the HTML. The HTML was directly served at the request of the client. The browser then downloaded the JavaScript, CSS, and image files according to the HTML. Most of the time, the JavaScript was only responsible for allowing dynamic content or layout changes, rather than rendering the entire application. This could include drop-down menus, accordions, or just pulling new data from the backend via **Ajax**. The main HTML of the application, however, was directly returned from the backend, which resulted in a monolith application. A significant benefit of this solution is that the client does not need to process all of the business logic, as it has already been done on the server.

However, when we talk about SSR in the context of React applications, we are referring to something different. At this point in the book, we have written a React application that renders on the client. We do not want to re-implement the rendering for the backend in a slightly different way. We also don't want to lose the ability to change data, pages, or the layout dynamically in the browser, as we already have an excellent working application with many interaction possibilities for the user.

An approach that allows us to make use of the pre-rendered HTML – as well as the dynamic features provided by React – is called **universal rendering**. With universal rendering, the first request of the client includes a pre-rendered HTML page. The HTML should be the exact HTML that the client generates when processing it on its own. If this is the case, React can reuse the HTML provided by the server. Since SSR not only involves reusing HTML but also saving requests made by Apollo, the client also needs a starting cache that React can rely on. The server makes all of the requests before sending the rendered HTML and inserts a state variable for Apollo and React into the HTML. The result is that on the first request by the client, our frontend should not need to rerender or refresh any HTML or data that is returned by the server. For all following actions, such as navigating to other pages or sending messages, the same client-side React code from before is used. In other words, SSR is only used on the first page load. Afterward, these features do not require SSR because the client-side code continues to work dynamically, as it did before.

Let's get started with writing some code.

Setting up SSR in Express.js to render React on the server

In this example, the first step is to implement basic SSR on the backend. We are going to extend this functionality later to validate the authentication of the user. An authenticated user allows us to execute Apollo or GraphQL requests rather than only render the pure React markup. First, we need some new packages. Because we are going to use universally rendered React code, we require an advanced webpack configuration. Therefore, we will install the following packages:

```
npm install --save-dev webpack-dev-middleware webpack-hot-
middleware @babel/cli
```

Let's quickly go through the packages that we are installing. We only need these packages for development:

- The first webpack module, called `webpack-dev-middleware`, allows the backend to serve bundles that are generated by webpack, but from memory and without creating files. This is convenient for cases in which we need to run JavaScript directly and do not want to use separate files.

- The second package, called `webpack-hot-middleware`, only handles client-side updates. If a new version of a bundle is created, the client is notified, and the bundle is exchanged.

- The last package, called `@babel/cli`, allows us to introduce the great features that **Babel** provides to our backend. We are going to use React code that has to be transpiled.

In a production environment, it is not recommended to use these packages. Instead, the bundle is built once before deploying the application. The client downloads the bundle when the application has gone live.

For development with SSR enabled, the backend uses these packages to distribute the bundled React code to the client after the SSR has finished. The server itself relies on the plain `src` files and not on the webpack bundle that the client receives.

We also depend on one further essential package, shown as follows:

```
npm install --save node-fetch
```

To set up **Apollo Client** on the backend, we require a replacement of the standard `window.fetch` method. Apollo Client uses this to send GraphQL requests, which is why we install `node-fetch` as a polyfill. We are going to set up Apollo Client for the backend later in this chapter.

Before starting with the primary work, ensure that your NODE_ENV environment variable is set to `development`.

Then, head over to the server's `index.js` file, where all of the Express.js magic happens. We didn't cover this file in the previous chapter because we are going to adjust it now to support SSR including the routing directly.

First, we will set up the development environment for SSR, as this is essential for our next tasks. Follow these steps to get your development environment ready for SSR:

1. The first step is to import two new webpack modules: `webpack-dev-middleware` and `webpack-hot-middleware`. These should only be used in a development environment, so we should require them conditionally by checking the environment variables. In a production environment, we generate the webpack bundles in advance. In order to only use the new packages in development, put the following code underneath the setup for the Express.js helmet:

```
if (process.env.NODE_ENV === 'development') {
    const devMiddleware =
      require('webpack-dev-middleware');
    const hotMiddleware =
      require('webpack-hot-middleware');
    const webpack = require('webpack');
    const config =
      require('../../webpack.server.config');
    const compiler = webpack(config);
    app.use(devMiddleware(compiler));
    app.use(hotMiddleware(compiler));
}
```

2. After loading these packages, we will also require webpack, as we will parse a new webpack configuration file. The new configuration file is only used for the SSR.

3. After both webpack and the configuration file have been loaded, we will use the `webpack(config)` command to parse the configuration and create a new webpack instance.

4. Next, we are going to create the webpack configuration file. To do this, we pass the created webpack instance to our two new modules. When a request reaches the server, the two packages take action according to the configuration file.

The new configuration file has only a few small differences as compared to the original configuration file, but these have a big impact. Create the new `webpack.server.config.js` file, and enter the following configuration:

```
const path = require('path');
const webpack = require('webpack');
const buildDirectory = 'dist';
```

```
module.exports = {
  mode: 'development',
  entry: [
    'webpack-hot-middleware/client',
    './src/client/index.js'
  ],
  output: {
    path: path.join(__dirname, buildDirectory),
    filename: 'bundle.js',
    publicPath: '/'
  },
  module: {
    rules: [
      {
        test: /\.js$/,
        exclude: /node_modules/,
        use: {
          loader: 'babel-loader',
        },
      },
      {
        test: /\.css$/,
        use: ['style-loader', 'css-loader'],
      },
      {
        test: /\.(png|woff|woff2|eot|ttf|svg)$/,
        loader: 'url-loader?limit=100000',

      },
    ],
  },
  plugins: [
    new webpack.HotModuleReplacementPlugin(),
    new webpack.NamedModulesPlugin(),
  ],
};
```

We have made three changes in the preceding configuration in comparison to the original `webpack.client.config.js` file, and these are as follows:

- In the `entry` property, we now have multiple entry points. The `index` file for the frontend code, as before, is one entry point. The second one is the new `webpack-hot-middleware` module, which initiates the connection between the client and the server. The connection is used to send the client notifications to update the bundle to a newer version.

- I removed the `devServer` field, as this configuration does not require webpack to start its own server. Express.js is the web server, which we are already using when loading the configuration.

- The plugins are entirely different from those of the client's webpack configuration. We do not need `CleanWebpackPlugin`, as this cleans the `dist` folder, nor the `HtmlWebpackPlugin` plugin, which inserts the webpack bundles into the `index.html` file; this is handled by the server differently. These plugins are only useful for client-side development. Now, we have `HotModuleReplacementPlugin`, which enables **Hot Module Replacement** (**HMR**). It allows for JavaScript and CSS to be exchanged on the fly. Next, `NamedModulesPlugin` displays the relative paths for modules injected by HMR. Both plugins are only recommended for use in development.

The webpack preparation is now finished.

Now, we have to focus on how to render React code and how to serve the generated HTML. However, we cannot use the existing React code that we have written. First, there are specific adjustments that we have to make to the main files: `index.js`, `App.js`, `router.js`, and `apollo/index.js`. Many packages that we use, such as **React Router** or Apollo Client, have default settings or modules that we have to configure differently when they are executed on the server.

We will begin with the root of our React application, which is the `index.js` file. We are going to implement an individual SSR `index` file, as there are server-specific adjustments to do.

Create a new folder called `ssr` inside the `server` folder. Then, insert the following code into an `index.js` file inside the `ssr` folder:

```
import React from 'react';
import { ApolloProvider } from '@apollo/client';
import App from './app';
```

```
const ServerClient = ({ client, location, context }) => {
  return (
    <ApolloProvider client={client}>
      <App location={location} context={context}/>
    </ApolloProvider>
  );
}


export default ServerClient
```

The preceding code is a modified version of our client index.js root file. The changes that the file has gone through are listed as follows:

- Instead of using the ReactDOM.render function to insert the HTML into the DOMNode with the root ID, we are now exporting a React component. The returned component is called ServerClient. There is no DOM that we can access to let ReactDOM render anything, so we skip this step when rendering on the server.

- The ApolloProvider component now receives Apollo Client directly from the ServerClient properties, whereas we previously set up Apollo Client directly inside this file by importing the index.js file from the apollo folder and passing it to the provider. You will soon see why we are doing this.

- The last change we made was to extract a location and a context property. We pass these properties to the App component. In the original version, there were no properties passed to the App component. Both properties are required in order to configure React Router to work with SSR. We are going to implement the properties later in the chapter.

Before looking at why we made these changes in more detail, let's create the new App component for the backend. Create an app.js file next to the index.js file in the ssr folder, and insert the following code:

```
import React, { useState } from 'react';
import { Helmet } from 'react-helmet';
import { withApollo } from '@apollo/client/react/hoc';
import Router from '../../client/router';
import { useCurrentUserQuery } from '../../client/apollo/
queries/currentUserQuery';
import '../../client/components/fontawesome';
```

```
const App = ({ location, context }) => {
  const { data, loading, error } = useCurrentUserQuery();
  const [loggedIn, setLoggedIn] = useState(false);

  return (
    <div className="container">
      <Helmet>
        <title>Graphbook - Feed</title>
        <meta name="description" content="Newsfeed of all
          your friends on Graphbook" />
      </Helmet>
      <Router loggedIn={loggedIn}
        changeLoginState={setLoggedIn} location={location}
          context={context} />
    </div>
  )
}

export default withApollo(App)
```

The following are a few changes that we made:

- The first change, in comparison to the original client-side App component, was to adjust the import statements to load the router and the fontawesome component from the client folder, as they do not exist in the server folder.

- The second change was to remove the useEffect Hooks and the localStorage access. We did this because the authentication that we built uses the localStorage access. This is fine for client-side authentication. Neither **Node.js** nor the server supports such storage, in general. Also, the useEffect Hook is only called on the client side. This is why we remove the authentication when moving our application to SSR. We are going to replace the localStorage implementation with cookies in a later step. For the moment, the user stays logged out of the server.

- The last change involves passing the two new properties, context and location, to the Router component in the preceding code.

React Router provides instant support for SSR. Nevertheless, we need to make some adjustments. The best way is that we use the same router for the backend and frontend, so we do not need to define routes twice, which is inefficient and can lead to problems. Open the `router.js` file inside the `client` folder and follow these steps:

1. Change the `import` statement for the `react-router-dom` package to look like the following:

   ```
   import { BrowserRouter, StaticRouter, Route, Redirect,
   Switch } from 'react-router-dom';
   ```

2. Insert the following code to extract the correct router:

   ```
   let Router;
   if(typeof window !== typeof undefined) {
     Router = BrowserRouter;
   }
   else {
     Router = StaticRouter;
   }
   ```

   After importing the React Router package, we check whether the file is executed on the server or the client by looking for the `window` object. Since there is no `window` object in Node.js, this is a sufficient check. An alternative approach would be to set up the `Switch` component, including the routes, in a separate file. This approach would allow us to import the routes directly into the correct router if we create two separate router files for client-side and server-side rendering.

   If we are on the client side, we use `BrowserRouter`, and if not, we use `StaticRouter`. Here, the logic is that with `StaticRouter`, we are in a stateless environment, where we render all routes with a fixed location. The `StaticRouter` component does not allow for the location to be changed by redirects because no user interaction can happen when using SSR. The other components, `Route`, `Redirect`, and `Switch`, can be used as before.

   Regardless of which routers are extracted, we save them in the `Router` variable. We then use them in the return statement of the `routing` component.

3. We prepared the `context` and `location` properties, which are passed from the top `ServerClient` component to the `Router` variable. If we are on the server, these properties should be filled, because the `StaticRouter` object requires them. You can replace the `Router` tag in the bottom `Routing` component, as follows:

    ```
    <Router context={this.props.context} location={this.
    props.location}>
    ```

    The `location` object holds the path that the router should render. The `context` variable stores all of the information the `Router` component processes, such as redirects. We can inspect this variable after rendering the `Router` component to trigger the redirects manually. This behavior is the big difference between `BrowserRouter` and `StaticRouter`. In the first case, `BrowserRouter` redirects the user automatically, but `StaticRouter` does not.

The crucial components to render our React code successfully have now been prepared. However, there are still some modules that we have to initialize before rendering anything with React. Open the `index.js` server file again. At the moment, we are serving the `dist` path statically on the root/path for **client-side rendering** (**CSR**), which can be found at `http://localhost:8000`. When moving to SSR, we have to serve the HTML generated by our React application at the `/` path instead.

Furthermore, any other path, such as `/app`, should also use SSR to render those paths on the server. Remove the current `app.get` method at the bottom of the file, which is right before the `app.listen` method. Then, insert the following code as a replacement:

```
app.use('/', express.static(path.join(root, 'dist/client'), {
index: false }));
app.get('*', (req, res) => {
  res.status(200);
  res.send('<!doctype html>');
  res.end();
});
```

The first line of the code should replace the old static route. It introduces a new option called `index`, which will disable serving the `index.html` file at the root path.

The asterisk (`*`) that we are using in the preceding code can overwrite any path that is defined later in the Express.js routing. Always remember that the `services` routine that we use in Express.js can implement new paths, such as `/graphql`, that we do not want to overwrite. To avoid this, put the code at the bottom of the file, below the `services` setup. The route catches any requests sent to the backend.

You can try out this route by running the `npm run server` command. Just visit `http://localhost:8000` to do this.

Currently, the preceding catch-all route only returns an empty site, with a status of `200`. Let's change this. The logical step would be to load and render the `ServerClient` component from the `index.js` file of the `ssr` folder, as it is the starting point of the React SSR code. The `ServerClient` component, however, requires an initialized Apollo Client instance, as we explained before. We are going to create a special Apollo Client instance for SSR next.

Create a `ssr/apollo.js` file, as it does not exist yet. We will set up Apollo Client in this file. The content is nearly the same as the original setup for the client:

```
import { ApolloClient } from 'apollo-client';
import { InMemoryCache } from 'apollo-cache-inmemory';
import { onError } from 'apollo-link-error';
import { ApolloLink } from 'apollo-link';
import { HttpLink } from 'apollo-link-http';
import fetch from 'node-fetch';

export default (req) => {
  const AuthLink = (operation, next) => {
    return next(operation);
  };
  const client = new ApolloClient({
    ssrMode: true,
    link: ApolloLink.from([
      onError(({ graphQLErrors, networkError }) => {
        if (graphQLErrors) {
          graphQLErrors.map(({ message, locations, path,
            extensions }) => {
            console.log('[GraphQL error]: Message:
              ${message},
                Location: ${locations}, Path: ${path}');
          });
          if (networkError) {
            console.log('[Network error]:
              ${networkError}');
```

```
          }
        }
      }),
      AuthLink,
      new HttpLink({
        uri: 'http://localhost:8000/graphql',
        credentials: 'same-origin',
        fetch
      })
    ]),
    cache: new InMemoryCache(),
  });
  return client;
};
```

However, there are a few changes that we made to get the client working on the server. These changes were pretty big, so we created a separate file for the server-side Apollo Client setup. Take a look at the changes (as follows) to understand the differences between the frontend and the SSR setup for Apollo Client:

- Instead of using the `createUploadLink` function that we introduced to allow the user to upload images or other files, we are now using the standard `HttpLink` class again. You could use the `UploadClient` function, but the functionalities it provides won't be used on the server, as the server won't upload files.

- The `AuthLink` function skips to the next link, as we have not implemented server-side authentication yet.

- The `HttpLink` object receives the `fetch` property, which is filled by the `node-fetch` package that we installed at the beginning of the chapter. This is used instead of the `window.fetch` method, which is not available in Node.js.

- Rather than exporting the `client` object directly, we export a wrapping function that accepts a `request` object. We pass it as a parameter in the Express.js route. As you can see in the preceding code example, we haven't used the object yet, but that will change soon.

Import the `ApolloClient` component at the top of the server `index.js` file, as follows:

```
import ApolloClient from './ssr/apollo';
```

The imported `ApolloClient` function accepts the `request` object of our Express.js server.

Add the following line to the top of the new Express.js catch-all route:

```
const client = ApolloClient(req);
```

This way, we set up a new `client` instance that we can hand over to our `ServerClient` component.

We can continue and implement the rendering of our `ServerClient` component. To make the future code work, we have to load React and, of course, the `ServerClient` component itself:

```
import React from 'react';
import Graphbook from './ssr/';
```

The `ServerClient` component is imported under the `Graphbook` name. We import React because we use the standard **JSX** syntax when rendering our React code.

Now that we have access to Apollo Client and the `ServerClient` component, insert the following two lines below the `ApolloClient` setup in the Express.js route:

```
const context= {};
const App = (<Graphbook client={client} location={req.url}
  context= {context}/>);
```

We pass the initialized `client` variable to the `Graphbook` component. We use the regular React syntax to pass all properties. Furthermore, we set the `location` property to the request object's `url` object, to tell the router which path to render. The `context` property is passed as an empty object.

However, why do we pass an empty object as `context` to the router at the end?

The reason is that after rendering the `Graphbook` component to HTML, we can access the `context` object and see whether a redirect (or something else) would have been triggered regularly. As we mentioned before, redirects have to be implemented by the backend code. The `StaticRouter` component of React Router does not make assumptions about the Node.js web server that you are using. That is why `StaticRouter` does not execute them automatically. Tracking and post-processing these events is possible with the `context` variable.

The resulting React object is saved to a new variable, which is called `App`. Now, there should be no errors if you start the server with `npm run server` and visit `http://localhost:8000`. Still, we see an empty page. That happens because we only return an empty HTML page; we haven't rendered the React `App` object to HTML. To render the object to HTML, import the following package at the top of the server `index.js` file:

```
import ReactDOM from 'react-dom/server';
```

The `react-dom` package not only provides bindings for the browser but also provides a special module for the server, which is why we use the `/server` suffix when importing it. The returned module provides a number of server-only functions.

> **Note**
>
> To learn some more advanced features of SSR and the dynamics behind it, you should read up on the official documentation of the server package of `react-dom` at `https://reactjs.org/docs/react-dom-server.html`.

We can translate the React `App` object into HTML by using the `ReactDOM.rendertoString` function. Insert the following line of code beneath the `App` object:

```
const content = ReactDOM.renderToString(App);
```

This function generates HTML and stores it inside the `content` variable. The HTML can be returned to the client now. If you return pre-rendered HTML from the server, the client goes through it and checks whether its current state would match the returned HTML. The comparison is made by identifying certain points in the HTML, such as the `data-reactroot` property.

If, at any point, the markup between the server-rendered HTML and that which the client would generate does not match, an error is thrown. The application will still work, but the client will not be able to make use of SSR; the client will replace the complete markup returned from the server by rendering everything again. In this case, the server's HTML response is thrown away. This is, of course, very inefficient and not what we are aiming for.

We have to return the rendered HTML to the client. The HTML that we have rendered begins with the root `div` tag and not the `html` tag. We must wrap the `content` variable inside a template that includes the surrounding HTML tags. So, create a `template.js` file inside the `ssr` folder, and enter the following code to implement the template for our rendered HTML:

```
import React from 'react';
import ReactDOM from 'react-dom/server';
```

```
const htmlTemplate = (content) => {
  return '
    <html lang="en">
      <head>
        <meta charSet="UTF-8"/>
        <meta name="viewport" content="width=device-width,
          initial-scale=1.0"/>
        <meta httpEquiv="X-UA-Compatible"
          content="ie=edge"/>
        <link rel="shortcut icon"
          href="data:image/x-icon;," type="image/x-icon">
        ${(process.env.NODE_ENV === 'development')? "":
          "<link rel='stylesheet' href='/bundle.css'/>"}
      </head>
      <body>
        ${ReactDOM.renderToStaticMarkup(<div id="root"
          dangerouslySetInnerHTML={{ __html: content
            }}></div>)}
        <script src="/bundle.js"></script>
      </body>
    </html>
  ';
};


export default htmlTemplate;
```

The preceding code is pretty much the same HTML markup as that in the index.html file that we usually serve to the client. The difference is that here, we use React and ReactDOM.

First, we export a function, which accepts the content variable with the rendered HTML.

Then, we render a link tag inside the head tag, which downloads the CSS bundle if we are in a production environment. For our current development scenario, there is no bundled CSS.

The important part is that we use a new ReactDOM function called rendertoStaticMarkup inside the body tag. This function inserts the React root tag into the body of our HTML template. Before, we used the renderToString method, which included special React tags, such as the data-reactroot property. Now, we use the rendertoStaticMarkup function to generate standard HTML without special React tags. The only parameter that we pass to the function is the div tag with the root ID and a new property, dangerouslySetInnerHTML. This attribute is a replacement for the regular innerHTML attribute, but for use in React. It lets React insert the HTML inside the root div tag. As the name suggests, it is dangerous to do this, but only if it is done on the client, as there is no possibility for **cross-site scripting (XSS)** attacks on the server. We use the ReactDOM.renderToStaticMarkup function to make use of the attribute. The inserted HTML was initially rendered with the renderToString function so that it would include all of the critical React HTML attributes and the wrapping div tag with the root ID. It can then be reused in the browser by the frontend code without any problems.

We require this template.js file in the server index file, at the top of the file:

```
import template from './ssr/template';
```

The template function can now be used directly in the res.send method, as follows:

```
res.send('<!doctype html>\n${template(content)}');
```

We do not only return a doctype object anymore; we also respond with the return value of the template function. As you should see, the template function accepts the rendered content variable as a parameter and composes it to a valid HTML document.

At this point, we have managed to get our first version of a server-side-rendered React application working. You can prove this by right-clicking in your browser window and choosing to view the source code. The window shows you the original HTML that is returned by the server. The output equals the HTML from the template function, including the login and signup forms.

Nevertheless, there are two problems that we face:

- There is no description meta head tag included in the server response. Something must have gone wrong with React Helmet.

- When logged in on the client side and, for example, viewing the news feed under the /app path, the server responds without having rendered the news feed or the login form. Normally, React Router would have redirected us to the login form since we are not logged in on the server side. However, since we use StaticRouter, we have to initiate the redirect separately, as we explained before. We are going to implement the authentication in a separate step.

We will start with the first issue. To fix the problem with React `Helmet`, import it at the top of the server `index.js` file, as follows:

```
import { Helmet } from 'react-helmet';
```

Now, before setting the response status with `res.status`, you can extract the React `Helmet` status, as follows:

```
const head = Helmet.renderStatic();
```

The `renderStatic` method is specially made for SSR. We can use it after having rendered the React application with the `renderToString` function. It gives us all of the head tags that would have been inserted throughout our code. Pass this `head` variable to the `template` function as a second parameter, as follows:

```
res.send('<!doctype html>\n${template(content, head)}');
```

Go back to the `template.js` file from the `ssr` folder. Add the `head` parameter to the exported function's signature. Then, add the following two new lines of code to the HTML's head tag:

```
${head.title.toString()}
${head.meta.toString()}
```

The `head` variable extracted from React `Helmet` holds a property for each `meta` tag. These tags provide a `toString` function that returns a valid HTML tag, which you can directly enter into the document's head object. The first problem should be fixed: all head tags are now inside the server's HTML response.

Let's focus on the second problem. The server response returns an empty React `root` tag when visiting a `PrivateRoute` component. As we explained previously, the reason for this is that the naturally initiated redirect does not get through to us, as we are using `StaticRouter`. We are redirected away from the `PrivateRoute` component because the authentication is not implemented for the server-side-rendered code. The first thing to fix is to handle the redirect, and we should also at least respond with the login form, instead of an empty React `root` tag. Later, we need to fix the authentication problem.

You would not notice this problem without viewing the source code of the server's response. The frontend downloads the `bundle.js` file and triggers the rendering on its own, as it knows about the authentication status of the user. The user would not notice that. Still, it is more efficient if the server sends the correct HTML directly. The HTML will be wrong if the user is logged in, but in the case of an unauthenticated user, the login form is pre-rendered by the server as it initiates the redirects.

To fix this issue, we can access the `context` object that has been filled by React Router after it has used the `renderToString` function. The final Express.js route should look like the following code example:

```
app.get('*', (req, res) => {
  const client = ApolloClient(req);
  const context= {};
  const App = (<Graphbook client={client}
    location={req.url} context= {context}/>);
  const content = ReactDOM.renderToString(App);
  if (context.url) {
    res.redirect(301, context.url);
  } else {
    const head = Helmet.renderStatic();
    res.status(200);
    res.send('<!doctype html>\n${template(content,
      head)}');
    res.end();
  }
});
```

The condition for rendering the correct route on the server is that we inspect the `context.url` property. If it is filled, we can initiate a redirect with Express.js. This will navigate the browser to the correct path. If the property is not filled, we can return the HTML generated by React.

This route renders the React code correctly up to the point at which authentication is required. The SSR route correctly renders all public routes, but none of the secure routes. That means that we only respond with the login form at the moment, as it is the only route that doesn't require authentication.

The next step is to implement authentication in connection with SSR to fix this huge issue.

Authentication with SSR

You should have noticed that we have removed most of the authentication logic from the server-side React code. The reason for this is that `localStorage` cannot be transmitted to the server on the initial loading of a page, which is the only case where SSR can be used at all. This leads to the problem that we cannot render the correct route because we cannot verify whether a user is logged in. The authentication has to be transitioned to cookies, which are sent with every request.

It is important to understand that cookies also introduce some security issues. We will continue to use the regular HTTP authorization header for the GraphQL API that we have written. If we use cookies for the GraphQL API, we will expose our application to potential **cross-site request forgery** (**CSRF**) attacks. The frontend code continues to send all GraphQL requests with the HTTP authorization header.

We will only use the cookies to verify the authentication status of a user and to initiate requests to our GraphQL API for the SSR of the React code. The SSR GraphQL requests will include the authorization cookie's value in the HTTP authorization header. Our GraphQL API only reads and verifies this header and does not accept cookies. As long as you do not mutate data when loading a page and only query for the data to render, there will be no security issues.

> **Tip**
> As the topic of CSRF and XSS is important, I recommend that you read up on it in order to fully understand how to protect yourself and your users. You can find a great article at `https://www.owasp.org/index.php/Cross-Site_Request_Forgery_(CSRF)`.

So, just follow these instructions to get authentcation on SSR running:

1. The first thing to do is install a new package with npm, as follows:

    ```
    npm install --save cookies
    ```

    The `cookies` package allows us to easily interact through the Express.js `request` object with the cookies sent by the browser. Instead of manually parsing and reading through the cookie string (which is just a comma-separated list), you can access the cookies with simple `get` and `set` methods. To get this package working, you have to initialize it inside Express.js.

2. Import the `cookies` and `jwt` packages, and also extract the `JWT_SECRET` string from the environment variables at the top of the server `index.js` file:

    ```
    import Cookies from 'cookies';
    import JWT from 'jsonwebtoken';
    const { JWT_SECRET } = process.env;
    ```

    To use the `cookies` package, we are going to set up a new middleware route.

3. Insert the following code before initializing the webpack modules and the
    services routine:

```
app.use(
  (req, res, next) => {
    const options = { keys: ['Some random keys'] };
    req.cookies = new Cookies(req, res, options);
    next();
  }
);
```

This new Express.js middleware initializes the cookies package under the req.
cookies property for every request that it processes. The first parameter of the
Cookies constructor is the request, the second is the response object, and the last
is an options parameter. This takes an array of keys with which the cookies are
signed. The keys are required if you want to sign your cookies for security reasons.
You should take care of this in a production environment. You can also specify a
secure property, which ensures that the cookies are only transmitted on secure
HTTPS connections.

4. We can now extract the authorization cookie and verify the authentication of
    the user. To do this, replace the beginning of the SSR route with the following code
    in the server's index.js file:

```
app.get('*', async (req, res) => {
  const token = req.cookies.get('authorization',
    { signed: true });
  var loggedIn;
  try {
    await JWT.verify(token, JWT_SECRET);
    loggedIn = true;
  } catch(e) {
    loggedIn = false;
  }
```

Here, I have added the async declaration to the callback function because we use
the await statement inside it. The second step is to extract the authorization
cookie from the request object with req.cookies.get. Importantly, we specify
the signed field in the options parameter, because only then will it successfully
return the signed cookies.

The extracted value represents the JWT that we generate when a user logs in. We can verify this with the typical approach that we implemented in *Chapter 6, Authentication with Apollo and React*. That is, we use the `await` statement when verifying the JWT. If an error is thrown, the user is not logged in. The state is saved in the `loggedIn` variable.

5. Pass the `loggedIn` variable to the `Graphbook` component, as follows:

```
const App = (<Graphbook client={client}
loggedIn={loggedIn} location={req.url}
context={context}/>);
```

Now, we can access the `loggedIn` property inside `index.js` from the `ssr` folder.

6. Extract the `loggedIn` state from the properties, and pass it to the `App` component in the `ssr` folder's `index.js` file, as follows:

```
<App location={location} context={context}
loggedIn={loggedIn}/>
```

Inside the `App` component, we do not need to set the `loggedIn` state directly to `false`, but we can take the property's value because it is determined before the `App` component is rendered. This flow is different from the client procedure, where the `loggedIn` state is determined inside the `App` component.

7. Change the `App` component in the `app.js` file in order to match the following code:

```
const App = ({ location, context, loggedIn: loggedInProp
}) => {
  const { data, loading, error } =
    useCurrentUserQuery();
  const [loggedIn, setLoggedIn] =
    useState(loggedInProp);
```

Here, the result is that we pass down the `loggedIn` value from our Express.js route, over the `Graphbook` and `App` components, to our `Router` component. This already accepts the `loggedIn` property in order to render the correct path for the user. At the moment, we still do not set the cookie on the backend when a user successfully logs in.

8. Open the `resolvers.js` file of our GraphQL server to fix that. We will change a few lines for the `login` and `signup` functions. Both resolver functions need the same changes, as both set the authentication token after login or signup. So, insert the following code directly above the return statement:

```
context.cookies.set(
    'authorization',
    token, { signed: true, expires: expirationDate,
        httpOnly: true, secure: false, sameSite: 'strict'
        }
);
```

The preceding function sets the cookies for the user's browser. The context object is only the Express.js `request` object where we have initialized the cookies package. The properties of the `cookies.set` function are pretty self-explanatory, but let's describe them as follows:

a. The `signed` field specifies whether the keys entered during the initialization of the `cookies` object should be used to sign the cookie's value.

b. The `expires` property takes a `date` object. It represents the time until which the cookie is valid. You can set the property to whatever date you want, but I would recommend a short period, such as one day. Insert the following code above the `context.cookies.set` statement in order to initialize the `expirationDate` variable correctly:

```
const cookieExpiration = 1;
const expirationDate = new Date();
expirationDate.setDate(
    expirationDate.getDate() + cookieExpiration
);
```

c. The `httpOnly` field secures the cookie so that it is not accessible by client-side JavaScript.

d. The `secure` property has the same meaning as it did when initializing the `Cookie` package. It restricts cookies to SSL connections only. This is a must when going to production, but it cannot be used when developing, as most developers develop locally, without an SSL certificate.

e. The `sameSite` field takes either `strict` or `lax` as a value. I recommend setting it to `strict`, as you want your GraphQL API or server to receive the cookie with every request, but you also want to exclude all cross-site requests, as this could be dangerous.

9. Now, we should clean up our code. Since we are using cookies, we can remove the `localStorage` authentication flow in the frontend code. Open the `App.js` file in the `client` folder. Remove the `componentWillMount` method, as we are reading from `localStorage` there.

The cookies are automatically sent with any request, and they do not need a separate binding as `localStorage` does. That also means that we need a special `logout` mutation that removes the cookie from the browser. JavaScript is not able to access or remove the cookie because we specified it as `httpOnly`. Only the server can delete it from the client.

10. Create a new `logout.js` file inside the `mutations` folder in order to create a `logout` mutation Hook. The content should look like the following:

```
import { gql, useMutation } from '@apollo/client';

export const LOGOUT = gql'
  mutation logout {
    logout {
      success
    }
  }
';

export const useLogoutMutation = () =>
useMutation(LOGOUT);
```

The preceding Hook only sends a simple `logout` mutation, without any parameters or further logic.

11. We should use the function inside the `logout.js` file of the `bar` folder in order to send the GraphQL request. Import the component at the top of the file, as follows:

```
import { useLogoutMutation } from '../../apollo/
mutations/logout';
```

12. Replace the `logout` method with the following code, in order to send the mutation upon clicking the **Logout** button:

```
const [logoutMutation] = useLogoutMutation();

const logout = () => {
  logoutMutation().then(() => {
    localStorage.removeItem('jwt');
    changeLoginState(false);
    client.stop();
    client.resetStore();
  });
}
```

Here, we have wrapped the original functions inside the call to the `logoutMutation` function. This sends the GraphQL request to our server.

13. To implement the mutation on the backend, add one line to the GraphQL `RootMutation` type, inside `schema.js`:

```
logout: Response @auth
```

It's required that the user that's trying to log out is authorized, so we use the `@auth` directive.

14. The corresponding resolver function is as follows. Add it to the `resolvers.js` file in the `RootMutation` property:

```
logout(root, params, context) {
  context.cookies.set(
    'authorization',
    '', { signed: true, expires: new Date(), httpOnly:
      true, secure: false, sameSite: 'strict' }
  );
  return {
    message: true
  };
},
```

The resolver function is minimal. It removes the cookie by setting the expiration date to the current time. This removes the cookie on the client when the browser receives the response because it is expired at that point. This behavior is an advantage in comparison to `localStorage`.

We have completed everything to make the authorization work with SSR. It is a very complex task, since authorization, SSR, and CSR have effects on the whole application. Every framework out there has its own approach to implementing this feature, so please take a look at them too.

If you look at the source code returned from our server after the rendering, you should see that the login form is returned correctly, as it was before. Furthermore, the server now recognizes whether the user is logged in. However, the server does not yet return the rendered news feed, the application bar, or the chats. Only a loading message is included in the returned HTML. The client-side code also does not recognize that the user is logged in. We will take a look at these problems in the next section.

Running Apollo queries with SSR

By nature, GraphQL queries via `HttpLink` are asynchronous. We have implemented a `loading` component to show the user a loading message while the data is being fetched.

This is the same thing that is happening while rendering our React code on the server. All of the routing is evaluated, including whether we are logged in. If the correct route is found, all GraphQL requests are sent. The problem is that the first rendering of React returns the loading state, which is sent to the client by our server. The server does not wait until the GraphQL queries are finished and it has received all of the responses to render our React code.

We will fix this problem now. The following is a list of things that we have to do:

- We need to implement authentication for the SSR Apollo Client instance. We already did this for the routing, but now, we need to pass the cookie to the server-side GraphQL request too.

- We need to use a React Apollo-specific method to render the React code asynchronously in order to wait for all responses of the GraphQL requests.

- Importantly, we need to return the Apollo cache state to the client. Otherwise, the client will re-fetch everything, as its state is empty upon the first load of the page.

Let's get started, as follows:

1. The first step is to pass the `loggedIn` variable from the Express.js SSR route to the `ApolloClient` function as a second parameter. Change the `ApolloClient` call inside the server's `index.js` file to the following line of code:

   ```
   const client = ApolloClient(req, loggedIn);
   ```

   Change the signature of the exported function from the `apollo.js` file to also include this second parameter.

2. Replace the `AuthLink` function inside the Apollo Client's setup for SSR with the following code:

   ```
   const AuthLink = (operation, next) => {
     if(loggedIn) {
       operation.setContext(context => ({
         ...context,
         headers: {
           ...context.headers,
           Authorization:
             req.cookies.get('authorization')
         },
       }));
     }
     return next(operation)
   };
   ```

   This `AuthLink` adds the cookies to the GraphQL requests by using the `request` object given by Express.js. The `request` object already holds the initialized cookies package, which we use to extract the authorization cookie. This only needs to be done if the user has been verified as logged in previously.

3. Import a new function from the Apollo package inside the server's `index.js` file. Replace the import of the `ReactDOM` package with the following line of code:

   ```
   import { renderToStringWithData } from "@apollo/client/
   react/ssr";
   ```

4. Originally, we used the `ReactDOM` server methods to render the React code to HTML. These functions are synchronous; that is why the GraphQL request did not finish. To wait for all GraphQL requests, replace all of the lines – beginning from the `rendertoString` function until the end of the SSR route inside the server's `index.js` file. The result should look as follows:

```
renderToStringWithData(App).then((content) => {
  if (context.url) {
    res.redirect(301, context.url);
  } else {
    const head = Helmet.renderStatic();
    res.status(200);
    res.send('<!doctype html>\n${template(content,
      head)}');
    res.end();
  }
});
```

The `renderToStringWithData` function renders the React code, including the data received by the Apollo requests. Since the method is asynchronous, we wrap the rest of our code inside a callback function.

Now, if you take a look at the HTML returned by your server, you should see the correct markup, including chats, images, and everything else. The problem is that the client does not know that all of the HTML is already there and can be reused. The client would re-render everything.

5. To let the client reuse the HTML that our server sends, we have to include Apollo Client's state with our response. Inside the preceding callback, access Apollo Client's state by inserting the following code:

```
const initialState = client.extract();
```

The `client.extract` method returns a big object, holding all cache information that the client has stored after using the `renderToStringWithData` function.

6. The state must be passed to the `template` function as a third parameter. So, change the `res.send` call to the following:

```
res.send('<!doctype html>\n${template(content, head,
initialState)}');
```

7. Inside the `template.js` file, extend the function declaration and append the `state` variable as a third parameter, after the `head` variable.

8. Insert the `state` variable, with the following line of code, inside the HTML body and above the `bundle.js` file. If you add it below the `bundle.js` file, it won't work correctly:

```
${ReactDOM.renderToStaticMarkup(<script
dangerouslySetInnerHTML=
{{__html: 'window.__APOLLO_STATE__=${JSON.
stringify(state).replace
(/</g, '\\u003c')}'}}/>)}
```

We use the `renderToStaticMarkup` function to insert another `script` tag. It sets a large, stringified JSON object as Apollo Client's starting cache value. The JSON object holds all of the results of the GraphQL requests returned while rendering our server-side React application. We directly store the JSON object as a string in a new field inside the `window` object. The `window` object is helpful since you can directly access the field globally.

9. Apollo has to know about the state variable. It can be used by Apollo Client to initialize its cache with the specified data, instead of having to send all of the GraphQL requests again. Open the `index.js` file from the client's `apollo` folder. The last property of the initialization process is the cache. We need to set our `__APOLLO_STATE__` instance as the starting value of the cache. Replace the `cache` property with the following code:

```
cache: new InMemoryCache().restore(window.__APOLLO_
STATE__)
```

We create the `InMemoryCache` instance and run its `restore` method, where we insert the value from the window object. Apollo Client should recreate its cache from this variable.

10. We have now set up the cache for Apollo. It will no longer run unnecessary requests for which the results already exist. Now, we can finally reuse the HTML, with one last change. We have to change `ReactDOM.render` to `ReactDOM.hydrate` in the client's `index.js` file. The difference between these functions is that React reuses the HTML if it was correctly rendered by our server. In this case, React only attaches some necessary event listeners. If you use the `ReactDOM.render` method, it dramatically slows down the initial rendering process, as it compares the initial DOM with the current DOM and may change it accordingly.

The last problem that we have is that the client-side code does not show the logged-in state of our application after refreshing a page. The server returns the correct markup with all the data, but the frontend redirects us to the login form. The reason for this is that we statically set the `loggedIn` state variable to `false` in the `App.js` file of the client-side code.

The best way to check whether the user is authenticated is to verify whether the `__APOLLO_STATE__` field on the window object is filled and has a `currentUser` object attached. If that is the case, we can assume that the user was able to fetch their own data record, so they must be logged in. To change our `App.js` file accordingly, add the following condition to the `loggedIn` state variable:

```
(typeof window.__APOLLO_STATE__ !== typeof undefined && typeof
window.__APOLLO_STATE__.ROOT_QUERY !== typeof undefined &&
typeof window.__APOLLO_STATE__.ROOT_QUERY.currentUser !==
typeof undefined)
```

As you can see in the preceding code, we verify whether the Apollo starting cache variable includes a `ROOT_QUERY` property with the `currentUser` subfield. The `ROOT_QUERY` property is filled if any query can be fetched successfully. The `currentUser` field is only filled if the authenticated user was successfully requested.

If you execute `npm run server`, you will see that everything now works perfectly. Take a look at the markup that's returned; you will see either the login form or, when logged in, all of the content of the page that you are visiting. You can log in on the client, the news feed is fetched dynamically, you can refresh the page, and all of the posts are directly there, without the need for a single GraphQL request, as the server returned the data side by side with the HTML. This works not only for the `/app` path but for any path that you implement.

We are now finished with the SSR setup.

So far, we have only looked at the developmental part of SSR. When we get to the point where we want to make a production build and publish our application, there are a few other things that we will have to consider, which we will look at in *Chapter 12, Continuous Deployment with CircleCI and AWS*.

Summary

In this chapter, we changed a lot of the code that we have programmed so far. You learned the advantages and disadvantages of offering SSR. The main principles behind React Router, Apollo, and authentication with cookies while using SSR should be clear by now. It takes a lot of work to get SSR running, and it needs to be managed with every change made to your application. Nevertheless, it has excellent performance and user experience benefits for your users.

In the next chapter, we will look at how to offer real-time updates through **Apollo Subscriptions**, instead of using the old and inefficient polling approach.

10
Real-Time Subscriptions

The GraphQL **application programming interface** (**API**) we have built is very advanced, as is the frontend. In the previous chapter, we introduced **server-side rendering** (**SSR**) to our application. We provided the user with a lot of information through the news feed, chats, and profile pages. The problem we are facing now, however, is that the user currently has to either refresh the browser or we have set a `pollInterval` property to Apollo Hooks to keep the display up to date. A better solution is to implement Apollo subscriptions through WebSockets. This allows us to refresh the **user interface** (**UI**) of the user with the newest user information in real time without manual user interaction or polling.

This chapter covers the following topics:

- Using GraphQL with WebSockets
- Implementing Apollo subscriptions
- JWT authentication with subscriptions
- Notifications with Apollo subscriptions

Technical requirements

The source code for this chapter is available in the following GitHub repository:

```
https://github.com/PacktPublishing/Full-Stack-Web-Development-
with-GraphQL-and-React-Second-Edition/tree/main/Chapter10
```

Using GraphQL with WebSockets

In *Chapter 1*, *Preparing Your Development Environment*, I explained all the main features that make GraphQL so useful. We mentioned that **HyperText Transfer Protocol** (**HTTP**) is the standard network protocol when using GraphQL. The problem with regular HTTP connections, however, is that they are one-time requests. They can only respond with the data that exists at the time of the request. If the database receives a change concerning the posts or the chats, the user won't know about this until they execute another request. The UI shows outdated data in this case.

To solve this issue, you can refetch all requests in a specific interval, but this is a bad solution because there's no time range that makes polling efficient. Every user would make unnecessary HTTP requests, which neither you nor the user wants.

The best solution relies on WebSockets instead of HTTP requests. As with HTTP, WebSockets are also based on the **Transmission Control Protocol** (**TCP**). One of the main features of WebSockets is that they allow bidirectional communication between the client and the server. Arguably, you could say that HTTP does the same since you send a request and get a response, but WebSockets work very differently. One requirement is that the web server supports WebSockets in general. If that's the case, the client can open a WebSocket connection to the server. The initial request to establish a WebSocket connection is a standard HTTP request. The server should then respond with a 101 status code. It tells the browser that it agrees to change the protocols from HTTP to WebSockets. If the connection is successful, the server can send updates through this connection to the client. These updates are also called messages or frames. There are no further requests needed by the client to let the server speak with the browser, unlike HTTP, where you always need a request first so that the server can respond to it.

Using WebSockets or Apollo subscriptions would fix the issue we encounter when using polling. We have one connection that stays open all the time. The server can send messages to the client whenever data is added or updated. WebSocket **Uniform Resource Locators** (**URLs**) start with `ws` or `wss` instead of the ordinary `http` or `https`. With WebSockets, you can also save valuable bandwidth for the users, but these are not included for WebSocket messages.

One disadvantage is that WebSockets are not a standard approach for implementing APIs. If you make your API public to third parties at some point, a standard HTTP API would likely fit better. Also, HTTP is much more optimized. HTTP requests can be cached and proxied easily with common web servers, such as nginx or Apache, but also by the browser itself, which is hard for WebSockets to do. The most significant impact on performance is that WebSocket connections are kept open as long as the user stays on your site. It's not a problem for one or a few hundred users, but scaling this to more people is likely to present you with some problems. However, it's still a very efficient solution to real-time web communication in contrast to polling, for example.

Most GraphQL client libraries are specialized and optimized for the standard HTTP protocol. It's the most common approach, so that's understandable. The people behind Apollo have got you covered; they've built-in support for WebSockets and for the implementation of GraphQL subscriptions. You can use those packages not only with Apollo but also with many other libraries. Let's get started with implementing Apollo subscriptions.

Digging deeper into Apollo subscriptions

With the earlier versions of Apollo Client, it was required that you install further packages to support WebSockets. Now, the only requirement is one further package that implements the WebSocket support on the server side.

> **Note**
>
> You can find an excellent overview and more details about Apollo subscriptions in the official documentation at `https://www.apollographql.com/docs/react/data/subscriptions/`.

The first step is to install all the required packages to get GraphQL subscriptions working. Install them using npm, as follows:

```
npm install --save subscriptions-transport-ws graphql-subscriptions
```

The following two packages provide the necessary modules for a subscription system:

- The `graphql-subscriptions` package provides the ability to connect our GraphQL backend with a **publish-subscribe** (**PubSub**) system. It gives the client the option to subscribe to specific channels and lets the backend publish new data to the client. It is an in-memory implementation that only works with one instance of our backend. It is discouraged for production use but it will help us to get it working locally.

- The `subscriptions-transport-ws` package gives our Apollo Server or other GraphQL libraries the option to accept WebSocket connections and accept queries, mutations, and subscriptions over WebSockets. Let's take a look at how we can implement subscriptions.

First, we are going to create a new subscription type next to the `RootQuery` and `RootMutation` types inside the GraphQL schema. You can set up events or entities that a client can subscribe to and receive updates inside the new subscription type. It only works by adding the matching resolver functions as well. Instead of returning real data for this new subscription type, you return a special object that allows the client to subscribe to events for the specific entity. These entities can be things such as notifications, new chat messages, or comments on a post. Each of them has got its own subscription channel.

The client can subscribe to these channels. It receives updates any time the backend sends a new WebSocket message—because data has been updated, for example. The backend calls a `publish` method that sends the new data through the subscription to all clients. You should be aware that not every user should receive all WebSocket messages since the content may include private data such as chat messages. There should be a filter before the update is sent to target only specific users. We'll see this feature later in the *Authentication with Apollo subscriptions* section.

Subscriptions on Apollo Server

We have now installed all the essential packages. Let's start with the implementation for the backend, as follows:

1. As mentioned previously, we are going to rely on WebSockets as they allow real-time communication between the frontend and the backend. We are first going to set up a new transport protocol for the backend.

   Open the `index.js` file of the server. Import a new Node.js interface at the top of the file, like this:

   ```
   import { createServer } from 'http';
   ```

   The `http` interface is included in Node.js by default. It handles the traditional HTTP protocol, making the use of many HTTP features easy for the developer.

2. We are going to use the interface to create a standardized Node.js HTTP `server` object because the Apollo `SubscriptionServer` module expects such an object. We'll cover the Apollo `SubscriptionServer` module soon in this section. Add the following line of code beneath the initialization of Express.js, inside the app variable:

   ```
   const server = createServer(app);
   ```

The createServer function creates a new HTTP server object, based on the original Express.js instance. We pass the Express instance, which we saved inside the app variable. As you can see in the preceding code snippet, you only pass the app object as a parameter to the createServer function.

3. We're going to use the new server object instead of the app variable to let our backend start listening for incoming requests. Remove the old app.listen function call from the bottom of the file because we'll be replacing it in a second. To get our server listening again, edit the initialization routine of the services. The for loop should now look like this:

```
for (let i = 0; i < serviceNames.length; i += 1) {
  const name = serviceNames[i];
  switch (name) {
    case 'graphql':
      (async () => {
        await services[name].start();
        app.use(graphqlUploadExpress());
        services[name].applyMiddleware({ app });
      })();
      break;
    case 'subscriptions':
      server.listen(8000, () => {
        console.log('Listening on port 8000!');
        services[name](server);
      });
      break;
    default:
      app.use('/${name}', services[name]);
      break;
  }
}
```

Here, we have changed the old if statement to a switch statement. Furthermore, we have added a second service beyond graphql, called subscriptions. We are going to create a new subscriptions service next to the graphql services folder.

The subscriptions service requires the server object as a parameter to start listening for WebSocket connections. Before initializing SubscriptionServer, we need to have started listening for incoming requests. That is why we use the server.listen method in the preceding code snippet before initializing the new subscriptions service that creates an Apollo SubscriptionServer instance. We pass the server object to the service after it has started listening. The service has to accept this parameter, of course, so keep this in mind.

4. To add the new service into the preceding serviceNames object, edit the index.js services file with the following content:

```
import graphql from './graphql';
import subscriptions from './subscriptions';

export default utils => ({
  graphql: graphql(utils),
  subscriptions: subscriptions(utils),
});
```

The subscriptions service also receives the utils object, as with the graphql service.

5. Now, create a subscriptions folder next to the graphql folder. To fulfill the import of the preceding subscriptions service, insert the service's index.js file into this folder. There, we can implement the subscriptions service. As a reminder, we pass the utils object and also the server object from before. The subscriptions service must accept two parameters in separate function calls.

6. If you have created a new subscription index.js file, import all the dependencies at the top of the file, as follows:

```
import { makeExecutableSchema } from '@graphql-tools/
schema';
import { SubscriptionServer } from 'subscriptions-
transport-ws';
import { execute, subscribe } from 'graphql';
import jwt from 'jsonwebtoken';
import Resolvers from '../graphql/resolvers';
import Schema from'../graphql/schema';
import auth from '../graphql/auth';
```

The preceding dependencies are almost the same as those that we are using for the `graphql` service, but we've added the `subscriptions-transport-ws` and `@graphql-tools/schema` packages. Furthermore, we've removed the `apollo-server-express` package. `SubscriptionServer` is the equivalent of `ApolloServer` but is used for WebSocket connections rather than HTTP. It usually makes sense to set up Apollo Server for HTTP and `SubscriptionServer` for WebSockets in the same file, as this saves us from processing `Schema` and `Resolvers` twice. It's easier to explain the implementation of subscriptions without the `ApolloServer` code in the same file, though. The last two things that are new in the preceding code snippet are the `execute` and `subscribe` functions that we import from the `graphql` package. You will see why we need these in the next section.

7. We begin the implementation of the new service by exporting a function with the `export default` statement and creating an `executableSchema` object (as you saw in *Chapter 2, Setting up GraphQL with Express.js*), as follows:

```
export default (utils) => (server) => {
  const executableSchema = makeExecutableSchema({
    typeDefs: Schema,
    resolvers: Resolvers.call(utils),
    schemaDirectives: {
      auth: auth
    },
  });
}
```

As you can see, we use the **ECMAScript 6 (ES6)** arrow notation to return two functions at the same time. The first one accepts the `utils` object and the second one accepts the `server` object that we create with the `createServer` function inside the `index.js` file of the server. This approach fixes the problem of passing two parameters in separate function calls. The schema is only created when both functions are called.

8. The second step is to start `SubscriptionServer` to accept WebSocket connections and, as a result, be able to use the GraphQL subscriptions. Insert the following code under `executableSchema`:

```
new SubscriptionServer({
  execute,
  subscribe,
```

```
    schema: executableSchema,
  },
  {
    server,
    path: '/subscriptions',
  });
```

We initialized a new `SubscriptionServer` instance in the preceding code. The first parameter we pass is a general `options` object for GraphQL and corresponds to the options of the `ApolloServer` class. The options are detailed as follows:

a. The `execute` property should receive a function that handles all the processing and execution of incoming GraphQL requests. The standard is to pass the `execute` function that we imported from the `graphql` package previously.

b. The `subscribe` property also accepts a function. This function has to take care of resolving a subscription to `asyncIterator`, which is no more than an asynchronous `for` loop. It allows the client to listen for execution results and reflect them to the user.

c. The last option we pass is the GraphQL schema. We do this in the same way as for `ApolloServer`.

The second parameter our new instance accepts is the `socketOptions` object. This holds settings to describe the way in which the WebSockets work, as outlined here:

d. The `server` field receives our `server` object, which we pass from the `index.js` file of the server as a result of the `createServer` function. `SubscriptionServer` then relies on the existing server.

e. The `path` field represents the endpoint under which the subscriptions are accessible. All subscriptions use the `/subscriptions` path.

> **Note**
>
> The official documentation for the `subscriptions-transport-ws` package offers a more advanced explanation of `SubscriptionServer`. Take a look to get an overview of all its functionalities: `https://github.com/apollographql/subscriptions-transport-ws#subscriptionserver`.

The client would be able to connect to the WebSocket endpoint at this point. There are currently no subscriptions, and the corresponding resolvers are set up in our GraphQL API.

9. Open the `schema.js` file to define our first subscription. Add a new type called `RootSubscription` next to the `RootQuery` and `RootMutation` types, including the new subscription called `messageAdded`, as follows:

```
type RootSubscription {
  messageAdded: Message
}
```

Currently, if a user sends a new message to another user, this isn't shown to the recipient right away.

The first option I showed you was to set an interval to request new messages. Our backend is now able to cover this scenario with subscriptions. The event or channel that the client can subscribe to is called `messageAdded`. We can also add further parameters, such as a chat **identifier (ID)**, to filter the WebSocket messages if necessary. When creating a new chat message, it is publicized through this channel.

10. We have added `RootSubscription`, but we need to extend the schema root tag too. Otherwise, the new `RootSubscription` type won't be used. Change the schema, as follows:

```
schema {
  query: RootQuery
  mutation: RootMutation
  subscription: RootSubscription
}
```

We have successfully configured the tree GraphQL main types. Next, we have to implement the corresponding resolver functions. Open the `resolvers.js` file and perform the following steps:

1. Import all dependencies that allow us to set up our GraphQL API with a `PubSub` system, as follows:

```
import { PubSub, withFilter } from 'graphql-
subscriptions';
const pubsub = new PubSub();
```

The `PubSub` system offered by the `graphql-subscriptions` package is a simple implementation based on the standard Node.js `EventEmitter` class. When going to production, it's recommended to use an external store, such as Redis, with this package.

2. We've already added the third `RootSubscription` type to the schema, but not the matching property on the `resolvers` object. The following code snippet includes the `messageAdded` subscription. Add it to the resolvers:

```
RootSubscription: {
  messageAdded: {
    subscribe: () =>
      pubsub.asyncIterator(['messageAdded']),
  }
},
```

The `messageAdded` property isn't a function but just a simple object. It contains a `subscribe` function that returns `AsyncIterable`. It allows our application to subscribe to the `messageAdded` channel by returning a promise that's only resolved when a new message is added. The next item that's returned is a promise, which is also only resolved when a message has been added. This method makes `asyncIterator` great for implementing subscriptions.

> **Note**
>
> You can learn more about how `asyncIterator` works by reading through the proposal at `https://github.com/tc39/proposal-async-iteration`.

3. When subscribing to the `messageAdded` subscription, there needs to be another method that publicizes the newly created message to all clients. The best location is the `addMessage` mutation where the new message is created. Replace the `addMessage` resolver function with the following code:

```
addMessage(root, { message }, context) {
  logger.log({
    level: 'info',
    message: 'Message was created',
  });
  return Message.create({
    ...message,
  }).then((newMessage) => {
    return Promise.all([
      newMessage.setUser(context.user.id),
      newMessage.setChat(message.chatId),
```

```
    ]).then(() => {
        pubsub.publish('messageAdded', {
          messageAdded: newMessage });
        return newMessage;
    });
  });
},
```

I have edited the `addMessage` mutation so that the correct user from the context is chosen. All of the new messages that you send are now saved with the correct user ID. This allows us to filter WebSocket messages for the correct users later in the *Authentication with Apollo subscriptions* section.

We use the `pubsub.publish` function to send a new WebSocket frame to all clients that are connected and that have subscribed to the `messageAdded` channel. The first parameter of the `pubsub.publish` function is the subscription, which in this case is `messageAdded`. The second parameter is the new message that we save to the database. All clients that have subscribed to the `messageAdded` subscription through `asyncIterator` now receive this message.

We've finished preparing the backend. The part that required the most work was to get the Express.js and WebSocket transport working together. The GraphQL implementation only involves the new schema entities, correctly implementing the resolver functions for the subscription, and then publishing the data to the client via the `PubSub` system.

We have to implement the subscription feature in the frontend to connect to our WebSocket endpoint.

Subscriptions on Apollo Client

As with the backend code, we also need to make adjustments to the Apollo Client configuration before using subscriptions. In *Chapter 4, Hooking Apollo into React*, we set up Apollo Client with the normal `HttpLink` link. Later, we exchanged it with the `createUploadLink` function, which enables the user to upload files through GraphQL.

We are going to extend Apollo Client by using `WebSocketLink` as well. This allows us to use subscriptions through GraphQL. Both links work side by side. We use the standard HTTP protocol to query data, such as the chat list or the news feed; all of these real-time updates to keep the UI up to date rely on WebSockets.

To configure Apollo Client correctly, follow these steps:

1. Open the `index.js` file from the `apollo` folder. Import the following dependencies:

    ```
    import { ApolloClient, InMemoryCache, from, split } from
    '@apollo/client';
    import { WebSocketLink } from '@apollo/client/link/ws';
    import { onError } from "@apollo/client/link/error";
    import { getMainDefinition } from '@apollo/client/
    utilities';
    import { createUploadLink } from 'apollo-upload-client';
    import { SubscriptionClient } from 'subscriptions-
    transport-ws';
    ```

    To get the subscriptions working, we need `SubscriptionClient`, which uses `WebSocketLink` to subscribe to our GraphQL API using WebSockets.

    We import the `getMainDefinition` function from the `@apollo/client/utilities` package. It's installed by default when using Apollo Client. The purpose of this function is to give you the operation type, which can be `query`, `mutation`, or `subscription`.

    The `split` function from the `@apollo/client` package allows us to conditionally control the flow of requests through different Apollo links based on the operation type or other information. It accepts one condition and one link (or a pair of links) from which it composes a single valid link that Apollo Client can use.

2. We are going to create both links for the `split` function. Detect the protocol and port where we send all GraphQL subscriptions and requests. Add the following code beneath the imports:

    ```
    const protocol = (location.protocol != 'https:') ?
    'ws://': 'wss://';
    const port = location.port ? ':'+location.port: '';
    ```

    The `protocol` variable saves the WebSocket protocol by detecting whether the client uses `http` or `https`. The `port` variable is either an empty string if we use port `80` to serve our frontend or any other port, such as `8000`, which we currently use. Previously, we had to statically save `http://localhost:8000` in this file. With the new variables, we can dynamically build the URL where all requests should be sent.

3. The `split` function expects two links to combine them into one. The first link is the normal `httpLink` link, which we must set up before passing the resulting link to the initialization of Apollo Client. Remove the `createUploadLink` function call from the `ApolloLink.from` function and add it before the `ApolloClient` class, as follows:

```
const httpLink = createUploadLink({
  uri: location.protocol + '//' + location.hostname +
    port + '/graphql',credentials: 'same-origin',
});
```

We concatenate the `protocol` variable of the server, which is either `http:` or `https:`, with two slashes. The `hostname` variable is, for example, the domain of your application or, if in development, `localhost`. The result of the concatenation is `http://localhost:8000/graphql`.

4. Add the WebSocket link that's used for the subscriptions next to `httpLink`. It's the second one we pass to the `split` function. The code is illustrated in the following snippet:

```
const SUBSCRIPTIONS_ENDPOINT = protocol +
  location.hostname + port + '/subscriptions';
const subClient =
  new SubscriptionClient(SUBSCRIPTIONS_ENDPOINT, {
  reconnect: true,
  connectionParams: () => {
    var token = localStorage.getItem('jwt');
    if(token) {
      return { authToken: token };
    }
    return { };
  }
});
const wsLink = new WebSocketLink(subClient);
```

We define the **Uniform Resource Identifier (URI)** that's stored inside the `SUBSCRIPTIONS_ENDPOINT` variable. It's built with the `protocol` and `port` variables, which we detected earlier, and the application's `hostname` variable. The URI ends with the specified endpoint of the backend with the same port as the GraphQL API. The URI is the first parameter of `SubscriptionsClient`. The second parameter allows us to pass options, such as the `reconnect` property. It tells the client to automatically reconnect to the backend's WebSocket endpoint when it has lost the connection. This usually happens if the client has temporarily lost their internet connection or the server has gone down.

Furthermore, we use the `connectionParams` field to specify the **JavaScript Object Notation (JSON) Web Token (JWT)** as an authorization token. We define this property as a function so that the JWT is read from `localStorage` whenever the user logs in. It's sent when the WebSocket is created.

We initialize `SubscriptionClient` to the `subClient` variable. We pass it to the `WebSocketLink` constructor under the `wsLink` variable with the given settings.

5. Combine both links into one. This allows us to insert the composed result into our `ApolloClient` class at the bottom. To do this, we have imported the `split` function. The syntax to combine the two links should look like this:

```
const link = split(
  ({ query }) => {
    const { kind, operation } =
      getMainDefinition(query);
    return kind === 'OperationDefinition' &&
      operation === 'subscription';
  },
  wsLink,
  httpLink,
);
```

The `split` function accepts three parameters. The first parameter must be a function with a Boolean return value. If the return value is `true`, the request is sent over the first link, which is the second required parameter. If the return value is `false`, the operation is sent over the second link, which we pass via the optional third parameter. In our case, the function that's passed as the first parameter determines the operation type. If the operation is a subscription, the function returns `true` and sends the operation over the WebSocket link. All other requests are sent via the HTTP Apollo link. We save the result of the `split` function in the `link` variable.

6. Insert the preceding `link` variable directly before the `onError` link. The `createUploadLink` function shouldn't be inside the `Apollo.from` function.

We've now got the basic Apollo Client set up to support subscriptions via WebSockets.

In *Chapter 5, Reusable React Components and React Hooks*, I gave you some homework to split the complete chat feature into multiple subcomponents. This way, the chat feature would follow the same pattern as we used for the post feed. We split it into multiple components so that it's a clean code base. We're going to use this and have a look at how to implement subscriptions for the chats.

If you haven't implemented the chat functionality in multiple subcomponents, you can get the working code from the official GitHub repository. I personally recommend you use the code from the repository if it's unclear what the following examples refer to.

Using chats as an example makes sense because they are, by nature, real time: they require the application to handle new messages and display them to the recipient. We take care of this in the following steps.

We begin with the main file of our chats feature, which is the `Chats.js` file in the client folder. I've reworked the `return` statement so that all the markup that initially came directly from this file is now entirely rendered by other child components. You can see all the changes in the following code snippet:

```
return (
  <div className="wrapper">
    <div className="chats">
      {chats.map((chat, i) =>
        <ChatItem chat={chat} user={user}
          openChat={openChat} />
      )}
    </div>
    <div className="openChats">
      {openChats.map((chatId, i) => <Chat chatId={chatId}
        key={"chatWindow" + chatId} closeChat={closeChat}
        /> )}
    </div>
  </div>
)
```

All the changes are listed here:

- We have introduced a new `ChatItem` component that handles the logic of the `for` loop. Extracting the logic into a separate file makes it more readable.

- The `ChatItem` component expects `user`, `chat`, and `openChat` properties. Furthermore, we have edited the functions that this component uses to also leverage the `user` object.

- We extract the `user` property from the properties of the `Chats` component. Consequently, we have to wrap the `Chats` component with the `UserConsumer` component to let it pass the user. You can apply this change from within the `Chats.js` file by wrapping the exported component into it.

- The `openChat` and `closeChat` functions are executed either by `ChatItem` or the `Chats` component. All other functions from the `Chats` component have been moved to one or both components: `ChatItem` and `Chat`.

The changes I have made here had nothing to do with the subscriptions directly, but it's much easier to understand what I'm trying to explain when the code is readable. If you need help implementing these changes by yourself, I recommend you check out the official GitHub repository. All the following examples are based on these changes, but they should be understandable without having the full source code.

More important, however, is `useGetChatsQuery`, which has a special feature. We want to subscribe to the `messageAdded` subscription to listen for new messages. That's possible by using a new function of the Apollo `useQuery` Hook.

We need to extract a `subscribeToMore` function from the `useGetChatsQuery` Hook.

The `subscribeToMore` function is provided by default with every result of an Apollo `useQuery` Hook. It lets you run an `update` function whenever a message is created. It works in the same way as the `fetchMore` function. We can use this function in the `Chats` component to listen for the new messages.

Let's have a look at how we can use this function to implement subscriptions on the frontend, as follows:

1. Create a new `subscriptions` folder inside the `apollo` folder.
2. Create a new `messageAdded.js` file inside this `subscriptions` folder. We need to parse the GraphQL subscription string. The new `messageAdded` subscription has to look like this:

```
export const MESSAGES_SUBSCRIPTION = gql'
  subscription onMessageAdded {
```

```
      messageAdded {
        id
        text
        chat {
          id
        }
        user {
          id
          __typename
        }
        __typename
      }
    }
  ';
```

The subscription looks exactly like all the other queries or mutations we are using. The only difference is that we request the __typename field, as it isn't included in the response of our GraphQL API when using subscriptions. From my point of view, this seems like a bug in the current version of SubscriptionServer. You should check whether you still need to do this at the time of reading this book.

We specify the operation type of the request, which is subscription, as you can see in the preceding code snippet. Otherwise, it attempts to execute the default query operation, which leads to an error because there's no messageAdded query, only a subscription. The subscription events the client receives when a new message is added hold all fields, as shown in the preceding code snippet.

3. In the addMessage mutation file, we need to rewrite one part of the code. We extract the fragment that we pass to writeFragment to be an exported variable itself so that we can reuse that. The code should look like this:

```
export const NEW_MESSAGE = gql'
  fragment NewMessage on Chat {
      id
      type
  }
';
```

4. Import the new GraphQL query in the `Chats.js` file together with some other dependencies, as follows:

```
import { MESSAGES_SUBSCRIPTION } from './apollo/queries/
messageAdded';
import { NEW_MESSAGE } from './apollo/mutations/
addMessage';
import { GET_CHAT } from './apollo/queries/getChat';
```

5. The following properties should be extracted from the `useGetChatsQuery` Hook:

```
const { loading, error, data, subscribeToMore } =
useGetChatsQuery();
```

6. Import the `withApollo` HOC and `UserConsumer`, as follows:

```
import { withApollo } from '@apollo/client/react/hoc';
import { UserConsumer } from './components/context/user';
```

7. We are going to use direct Apollo Client interaction. This is why we need to export the `Chats` component to be wrapped into the `withApollo` HOC to pass the client into a property. To export your component correctly, use the `withApollo` HOC. The code is illustrated here:

```
const ChatContainer = (props) => <UserConsumer><Chats
{...props} /></UserConsumer>

export default withApollo(ChatContainer)
```

We wrap the `Chats` component into the `UserConsumer` component to get access to the client. Furthermore, we wrap it into the `withApollo` HOC to get access to the client.

8. Here's the crucial part. When the component is mounted, we need to subscribe to the `messageAdded` channel. Only then is the `messageAdded` subscription used to receive new data or, to be exact, new chat messages. To start subscribing to the GraphQL subscription, we have to add a new `useEffect` Hook, as follows:

```
useEffect(() => {
  subscribeToNewMessages()
}, []);
```

In the preceding code snippet, we execute a new `subscribeToNewMessages` method inside the `useEffect` Hook of our React component.

The `useEffect` method only executes on the client-side code as the SSR implementation doesn't throw this event.

We have to add the corresponding `subscribeToNewMessages` method as well. We're going to explain every bit of this function in a moment. Insert the following code into the `Chats` component:

```
const subscribeToNewMessages = () => {
  subscribeToMore({
    document: MESSAGES_SUBSCRIPTION,
    updateQuery: (prev, { subscriptionData }) => {
      if (!subscriptionData.data || (prev.chats &&
          !prev.chats.length)) return prev;

      var index = -1;
      for(var i = 0; i < prev.chats.length; i++) {
        if(prev.chats[i].id ==
          subscriptionData.data.messageAdded.chat.id) {
          index = i;
          break;
        }
      }

      if (index === -1) return prev;

      const newValue = Object.assign({},prev.chats[i], {
        lastMessage: {
          text: subscriptionData.data.messageAdded.text,
          __typename:
            subscriptionData.data.messageAdded.__typename
        }
      });

      var newList = {chats:[...prev.chats]};
      newList.chats[i] = newValue;
```

```
        return newList;
      }
    });
  }
```

The preceding `subscribeToNewMessages` method looks very complex, but once we understand its purpose, it's straightforward. We primarily rely on the `subscribeToMore` function here, which we get from `useGetChatsQuery`. The purpose of this function is to start subscribing to our `messageAdded` channel and to accept the new data from the subscription and merge it with the current state and cache so that it's reflected directly to the user.

The `document` parameter accepts the parsed GraphQL subscription.

The second parameter is called `updateQuery`. It allows us to insert a function that implements the logic to update the Apollo Client cache with the new data. This function needs to accept a new parameter, which is the previous data from where the `subscribeToMore` function has been passed. In our case, this object contains an array of chats that already exist in the client's cache.

The second parameter holds the new message inside the `subscriptionData` index. The `subscriptionData` object has a `data` property that has a further `messageAdded` field under which the real message that's been created is saved.

We'll quickly go through the logic of the `updateQuery` function so that you can understand how we merge data from a subscription to the application state.

If `subscriptionData.data` is empty or there are no previous chats in the `prev` object, there's nothing to update. In this case, we return the previous data because a message was sent in a chat that the client doesn't have in their cache. Otherwise, we loop through all the previous chats of the `prev` object and find the index of the chat for which the subscription has returned a new message by comparing the chat IDs. The found chat's index inside the `prev.chats` array is saved in the `index` variable. If the chat cannot be found, we can check this with the `index variable` and return the previous data. If we find the chat, we need to update it with a new message. To do this, we compose the chat from the previous data and set `lastMessage` to the new message's text. We do this by using the `Object.assign` function, where the chat and the new message are merged. We save the result in the `newValue` variable. It's important that we also set the returned `__typename` property because otherwise, Apollo Client throws an error.

Now that we have an object that contains the updated chat in the newValue variable, we write it to the client's cache. To write the updated chat to the cache, we return an array of all chats at the end of the updateQuery function. Because the prev variable is read-only, we can't save the updated chat inside it. We have to create a new array to write it to the cache. We set the newValue chat object to the newList array at the index where we found the original chat. At the end, we return the newList variable. We update the cache that's given to us inside the prev object with the new array. Importantly, the new cache has to have the same fields as before. The schema of the return value of the updateQuery function must match the initial chats query schema.

You can now test the subscription directly in your browser by starting the application with npm run server. If you send a new chat message, it's shown directly in the chat panel on the right-hand side.

We have, however, got one major problem. If you test this with a second user, you'll notice that the lastMessage field is updated for both users. That is correct, but the new message isn't visible inside the chat window for the recipient. We've updated the client store for the chats GraphQL request, but we haven't added the message to the single chat query that's executed when we open a chat window.

We're going to solve this problem by making use of the withApollo HOC. The Chats component has no access to the chat query cache directly. The withApollo HOC gives the exported component a client property, which allows us to interact directly with Apollo Client. We can use it to read and write to the whole Apollo Client cache, and it isn't limited to only one GraphQL request. Before returning the updated chats array from the updateQuery function, we have to read the state of chat and insert the new data if possible. Insert the following code right before the final return statement inside the updateQuery function:

```
if (user.id !== subscriptionData.data.messageAdded.user.id) {
  try {
    const data = client.readQuery({ query: GET_CHAT,
      variables: { chatId:
        subscriptionData.data.messageAdded.chat.id } });
    client.cache.modify({
      id: client.cache.identify(data.chat),
      fields: {
        messages(existingMessages = []) {
          const newMessageRef =
            client.cache.writeFragment({
```

```
            data: subscriptionData.data.messageAdded,
            fragment: NEW_MESSAGE
          });
          return [...existingMessages, newMessageRef];
        }
      }
    });
  } catch(e) {}
}
```

In the preceding code snippet, we use the `client.readQuery` method to read the cache. This accepts the `GET_CHAT` query as one parameter and the chat ID of the newly sent message to get a single chat in return. The `GET_CHAT` query is the same request we sent in the `Chat.js` file when opening a chat window. We wrap the `readQuery` function in a `try-catch` block because it throws an unhandled error if nothing is found for the specified `query` and `variables`. This can happen if the user hasn't opened a chat window yet, and so no data has been requested with the `GET_CHAT` query for this specific chat. Furthermore, the whole block is wrapped into an `if` condition to check if the new message is from another user and not from ourselves because if we send a message on our own, we do not need to add it into the cache as we already do that on submission of a new message from our side.

If the message is from another user, we use the `client.cache.modify` function, as we already know to add the new message to the array of messages in the cache for this specific chat.

You can test these new changes by viewing the chat window and sending a message from another user account. The new message should appear almost directly for you without the need to refresh the browser.

In this section, we learned how to subscribe to events sent from a backend through Apollo subscriptions. Currently, we use this feature to update the UI on the fly with the new data. Later, in the *Notifications with Apollo subscriptions* section, we'll see another scenario where subscriptions can be useful. Nevertheless, there's one thing left to do: we haven't authorized the user for the `messageAdded` subscription through a JWT, such as our GraphQL API, and still, the user received the new message without verifying its identity. We're going to change this in the next section.

Authentication with Apollo subscriptions

In *Chapter 6, Authentication with Apollo and React*, we implemented authentication through the local storage of your browser. The backend generates a signed JWT that the client sends with every request inside the HTTP headers. In *Chapter 9, Implementing Server-Side Rendering*, we extended this logic to support cookies to allow SSR. Now that we've introduced WebSockets, we need to take care of them separately, as we did with the SRR and our GraphQL API.

How is it possible for the user to receive new messages when they aren't authenticated on the backend for the WebSocket transport protocol?

The best way to figure this out is to have a look at your browser's developer tools. Let's assume that we have one browser window where we log in with user A. This user chats with another user, B. Both send messages to each other and receive the new updates directly in their chat window. Another user, C, shouldn't be able to receive any of the WebSocket updates. We should play through this scenario in reality.

If you use Chrome as your default browser, go to the **Network** tab. There, you can filter all network requests by type. Since the data is transported via a WebSocket, you can filter by the **WS** option. You should see one connection, which is the subscriptions endpoint of our backend.

Try this scenario with the Developer Tools open. You should see the same WebSocket frames for all browsers. It should look like this:

Figure 10.1 – WebSocket messages

In the left panel, you can see all WebSocket connections. In our case, this is only the `subscriptions` connection. If you click on the connection, you will find all the frames that are sent over this connection. The first frame in the preceding list is the initial connection frame. The second frame is the subscription request to the `messageAdded` channel, which is initiated by the client. Both frames are marked green because the client sends them.

The last two are marked in red as the server sent them. The first of the red-marked frames is the server's acknowledgment of the established connection. The last frame was sent by our backend to publish a new message to the client. While the frame might look alright at first glance, it represents a vital problem. The last frame was sent to all clients, not just those who are members of the specific chat in which the message was sent. Average users are not likely to notice it since our `cache.modify` function only updates the UI if the chat was found in the client store. Still, an experienced user or developer is able to spy on all users of our social network as it's readable in the **Network** tab.

We need to take a look at the backend code that we have written and compare the initialization of `ApolloServer` and `SubscriptionServer`. We have a `context` function for `ApolloServer` that extracts the user from the JWT. It can then be used inside the resolver functions to filter the results by the currently logged-in user. For `SubscriptionServer`, there's no such `context` function at the moment. We have to know the currently logged-in user to filter the subscription messages for the correct users. We can use standard WebSockets events, such as `onConnect` or `onOperation`, to implement the authorization of the user.

The `onOperation` function is executed for every WebSocket frame that is sent. The best approach is to implement the authorization in the `onConnect` event in the same way as the `context` function that's taken from `ApolloServer` so that the WebSocket connection is authenticated only once when it's established and not for every frame that's sent.

In `index.js`, from the `subscriptions` folder of the server, add the following code to the first parameter of the `SubscriptionServer` initialization. It accepts an `onConnect` parameter as a function, which is executed whenever a client tries to connect to the `subscriptions` endpoint. Add the code just before the `schema` parameter:

```
onConnect: async (params, socket) => {
  const authorization = params.authToken;
  if(typeof authorization !== typeof undefined) {
    var search = "Bearer";
    var regEx = new RegExp(search, "ig");
```

```
    const token = authorization.replace(regEx, '').trim();
    return jwt.verify(token, JWT_SECRET, function(err,
      result) {
      if(err) {
        throw new Error('Missing auth token!');
      } else {
        return utils.db.models.User.findByPk(
          result.id).then((user) =>
          {
            return Object.assign({}, socket.upgradeReq, {
              user });
          });
        }
      });
    } else {
      throw new Error('Missing auth token!');
    }
  },
```

This code is very similar to the context function. We rely on the normal JWT authentication but via the connection parameters of the WebSocket. We implement the WebSocket authentication inside the onConnect event. In the original context function of ApolloServer, we extract the JWT from the HTTP headers of the request, but here, we are using the params variable, which is passed as the first parameter.

Before the client finally connects to the WebSocket endpoint, an onConnect event is triggered where you can implement special logic for the initial connection. With the first request, we send the JWT because we have configured Apollo Client to read the JWT to the authToken parameter of the connectionParams object when SubscriptionClient is initialized. That's why we can access the JWT not from a request object directly but from params.authToken in the preceding code snippet. The socket parameter is also given to us inside the onConnect function; there, you can access the initial upgrade request inside the socket object. After extracting the JWT from the connection parameters, we can verify it and authenticate the user with that.

At the end of this `onConnect` function, we return the `upgradeReq` variable and the user, just as we do with a normal `context` function for Apollo Server. Instead of returning the `req` object to `context` if the user isn't logged in, we are now throwing an error. This is because we only implement subscriptions for entities that require you to be logged in, such as chats or posts. It lets the client try to reconnect until it's authenticated. You can change this behavior to match your needs and let the user connect to the WebSocket. Don't forget, however, that every open connection costs you performance and a user who isn't logged in doesn't need an open connection, at least for the use case of **Graphbook**.

We have now identified the user that has connected to our backend with the preceding code, but we're still sending every frame to all users. This is a problem with the resolver functions because they don't use the context yet. Replace the `messageAdded` subscription with the following code in the `resolvers.js` file:

```
messageAdded: {
  subscribe: withFilter(() =>
    pubsub.asyncIterator('messageAdded'),
    (payload, variables, context) => {
    if (payload.messageAdded.UserId !== context.user.id) {
      return Chat.findOne({
        where: {
          id: payload.messageAdded.ChatId
        },
        include: [{
          model: User,
          required: true,
          through: { where: { userId: context.user.id } },
        }],
      }).then((chat) => {
        if(chat !== null) {
          return true;
        }
        return false;
      })
    }
    return false;
  }),
}
```

Earlier in this chapter, we imported the `withFilter` function from the `graphql-subscriptions` package. This allows us to wrap `asyncIterator` with a filter. The purpose of this filter is to conditionally send publications through connections to users who should see the new information. If one user shouldn't receive a publication, the return value of the condition for the `withFilter` function should be `false`. For all users who should receive a new message, the return value should be `true`.

The `withFilter` function accepts `asyncIterator` as its first parameter. The second parameter is the function that decides whether a user receives a subscription update. We extract the following properties from the function call:

- The `payload` parameter, which is the new message that has been sent in the `addMessage` mutation.

- The `variables` field, which holds all GraphQL parameters that could be sent with the `messageAdded` subscription, not with the mutation. For our scenario, we are not sending any variables with the subscription.

- The `context` variable, which holds all the information that we implemented in the `onConnect` Hook. It includes the regular `context` object with the user as a separate property.

The `filter` function is executed for every user that has subscribed to the `messageAdded` channel. First, we check whether the user for which the function is executed is the author of the new message by comparing the user IDs. In this case, they don't need to get a subscription notification because they already have the data.

If this isn't the case, we query the database for the chat where the new message was added. To find out whether a user needs to receive the new message, we select only chats where the logged-in user's ID and the chat ID are included. If a chat is found in the database, the user should see the new message. Otherwise, they aren't allowed to get the new message, and we return `false`.

Remember that the `withFilter` function is run for each connection. If there are thousands of users, we would have to run the database query very frequently. It's better to keep such filter functions as small and efficient as possible. For example, we could query the chat once to get the attached users and loop through them manually for all the connections. This solution would save us expensive database operations.

This is all you need to know about authentication with subscriptions. We now have a working setup that includes SSR with cookies and real-time subscriptions with JWT authentication. The SSR doesn't implement subscriptions because it doesn't make sense to offer real-time updates for the initial rendering of our application. Next, you will see another scenario where Apollo subscriptions can be useful.

Notifications with Apollo subscriptions

In this section, I'll quickly guide you through the second use case for subscriptions. Showing notifications to a user is perfectly traditional and commonplace, as you know from Facebook. Instead of relying on the subscribeToMore function, we use the Subscription component that's provided by Apollo. This component works like the Query and Mutation components, but for subscriptions.

Follow these steps to get your first Subscription component running:

1. Create a subscriptions folder inside the client's apollo folder. You can save all subscriptions that you implement using Apollo's useSubscription Hook inside this folder.

2. Insert a messageAdded.js file into the folder and paste in the following code:

    ```
    import { useSubscription, gql } from '@apollo/client';

    export const MESSAGES_SUBSCRIPTION = gql'
      subscription onMessageAdded {
        messageAdded {
          id
          text
          chat {
            id
          }
          user {
            id
            __typename
          }
          __typename
        }
      }
    ';

    export const useMessageAddedSubscription = (options) =>
    useSubscription(MESSAGES_SUBSCRIPTION, options);
    ```

The general workflow for the useSubscription component is the same as for the useMutation and useQuery Hooks. First, we parse the subscription with the gql function. Then, we just return the useSubscription Hook with the parsed GraphQL query.

3. Because we want to show notifications to the user when a new message is received, we install a package that takes care of showing pop-up notifications. Install it using npm, as follows:

```
npm install --save react-toastify
```

4. To set up react-toastify, add a ToastContainer component to a global point of the application where all notifications are rendered. This container isn't only used for notifications of new messages but for all notifications, so choose wisely. I decided to attach ToastContainer to the Chats.js file. Import the dependency at the top of it, as follows:

```
import { ToastContainer, toast } from 'react-toastify';
```

5. Inside the return statement, the first thing to render should be ToastContainer. Add it, like this:

```
<div className="wrapper">
  <ToastContainer/>
```

6. In the Chats.js file, add one import statement to load the subscription Hook, as follows:

```
import { useMessageAddedSubscription } from './apollo/
subscriptions/messageAdded';
```

7. Then, just call this subscription Hook inside the Chats component after the other Hook statements, like this:

```
useMessageAddedSubscription({
  onSubscriptionData: data => {
    if(data && data.subscriptionData &&
      data.subscriptionData.data &&
      data.subscriptionData.data.messageAdded)
      toast(data.subscriptionData.data.messageAdded.text,
        { position: toast.POSITION.TOP_LEFT });
  }
});
```

8. Add a small **Cascading Style Sheets** (**CSS**) rule and import the CSS rules of the `react-toastify` package. Import the CSS file in the `App.js` file, like this:

    ```
    import 'react-toastify/dist/ReactToastify.css';
    ```

    Then, add these few lines to the custom `style.css` file:

    ```
    .Toastify__toast-container--top-left {
      top: 4em !important;
    }
    ```

You can see an example of a notification in the following screenshot:

Figure 10.2 – Notification

The entire subscriptions topic is complex, but we managed to implement it for two use cases and thus provided the user with significant improvements to our application.

Summary

This chapter aimed to offer the user a real-time UI that allows them to chat comfortably with other users. We also looked at how to make this UI extendable. You learned how to set up subscriptions with any Apollo or GraphQL backend for all entities. We also implemented WebSocket-specific authentication to filter publications so that they are only received by the correct user.

In the next chapter, you'll learn how to verify and test the correct functionality of your application by implementing automated testing for your code.

11

Writing Tests for React and Node.js

So far, we've written a lot of code and come across a variety of problems. We haven't yet implemented automated testing for our software; however, it's a common approach to make sure everything works after making changes to your application. Automated testing drastically improves the quality of your software and reduces errors in production.

To achieve this, we will cover the following main topics in the chapter:

- How to use Mocha for testing

- Testing the GraphQL **application programming interface** (**API**) with Mocha and Chai

- Testing React with Enzyme and JSDOM

Technical requirements

The source code for this chapter is available in the following GitHub repository:

```
https://github.com/PacktPublishing/Full-Stack-Web-Development-
with-GraphQL-and-React-Second-Edition/tree/main/Chapter11
```

Testing with Mocha

The problem we're facing is that we must ensure the quality of our software without increasing the amount of manual testing. It isn't possible to recheck every feature of our software when new updates are released. To solve this problem, we're going to use Mocha, which is a JavaScript testing framework that is used to run a series of asynchronous tests. If all the tests pass successfully, your application is OK and can get released to production.

Many developers follow the **test-driven development** (**TDD**) approach. Often, when you implement tests for the first time, they fail because the business logic that's being tested is missing. After implementing all the tests, we have to write the actual application code to meet the requirements of the tests. In this book, we haven't followed this approach, but it isn't a problem as we can implement tests afterward too. Typically, I tend to write tests in parallel with the application code.

To get started, we have to install all the dependencies to test our application with npm, as follows:

```
npm install --save-dev mocha chai @babel/polyfill request
```

The mocha package includes almost everything to run tests. Along with Mocha, we also install chai, which is an assertion library. It offers excellent ways to chain tests with many variables and types for use inside a Mocha test. We also install the @babel/ polyfill package, which allows our test to support **ECMAScript 2015+** (**ES2015+**) syntax. This package is crucial because we use this syntax everywhere throughout our React code. Finally, we install the request package as a library to send all the queries or mutations within our test. I recommend you set the NODE_ENV environment variable to production to test every functionality, as in a live environment. Be sure that you set the environment variable correctly so that all production features are used.

Our first Mocha test

First, let's add a new command to the scripts field of our package.json file, as follows:

```
"test": "mocha --exit test/ --require babel-hook --require @
babel/polyfill --recursive"
```

If you now execute npm run test, we'll run the mocha package in the test folder, which we'll create in a second. The preceding --require option loads the specified file or package. We'll also load a babel-hook.js file, which we'll create as well. The --recursive parameter tells Mocha to run through the complete file tree of the test folder, not just the first layer. This behavior is useful because it allows us to structure our tests in multiple files and folders.

Let's begin with the `babel-hook.js` file by adding it to the root of our project, next to the `package.json` file. Insert the following code:

```
require("@babel/register")({
  "plugins": [
    "require-context-hook"
  ],
  "presets": ["@babel/env","@babel/react"]
});
```

The purpose of this file is to give us an alternative Babel configuration file to our standard `.babelrc` file. If you compare both files, you should see that we use the `require-context-hook` plugin. We already use this plugin when starting the backend with `npm run server`. It allows us to import our Sequelize models using a **regular expression (regex)**.

If we start our test with `npm run test`, we require this file at the beginning. Inside the `babel-hook.js` file, we load `@babel/register`, which compiles all the files that are imported afterward in our test according to the preceding configuration.

> **Note**
>
> Notice that when running a production build or environment, the production database is also used. All changes are made to this database. Verify that you have configured the database credentials correctly in the server's `configuration` folder. You have only to set the `host`, `username`, `password`, and `database` environment variables correctly.

This gives us the option to start our backend server from within our test file and render our application on the server. The preparation for our test is now finished. Create a folder named `test` inside the root of our project to hold all runnable tests. Mocha will scan all files or folders, and all tests will be executed. To get a basic test running, create an `app.test.js` file. This is the main file, which makes sure that our backend is running and in which we can subsequently define further tests. The first version of our test looks like this:

```
const assert = require('assert');
const request = require('request');
const expect = require('chai').expect;
const should = require('chai').should();
```

```
describe('Graphbook application test', function() {

    it('renders and serves the index page', function(done) {
        request('http://localhost:8000', function(err, res,
            body) {
            should.not.exist(err);
            should.exist(res);
            expect(res.statusCode).to.be.equal(200);
            assert.ok(body.indexOf('<html') !== -1);
            done(err);
        });
    });

});
```

Let's take a closer look at what's happening here, as follows:

1. We import the Node.js `assert` function. This gives us the ability to verify the value or type of a variable.

2. We import the `request` package, which we use to send queries against our backend.

3. We import two Chai functions, `expect` and `should`, from the `chai` package. Neither of these is included in Mocha, but they both improve the test's functionality significantly.

4. The beginning of the test starts with the `describe` function. Because Mocha executes the `app.test.js` file, we're in the correct scope and can use all Mocha functions. The `describe` function is used to structure your test and its output.

5. We use the `it` function, which initiates the first test.

The `it` function can be understood as a feature of our application that we want to test inside the callback function. As the first parameter, you should enter a sentence, such as `'it does this and that'`, that's easily readable. The function itself waits for the complete execution of the `callback` function in the second parameter. The result of the callback will either be that all assertions were successful or that, for some reason, a test failed or the callback didn't complete in a reasonable amount of time.

The `describe` function is the header of our test's output. Then, we have a new row for each `it` function we execute. Each row represents a single test step. The `it` function passes a `done` function to the callback. The `done` function has to be executed once all assertions are finished and there's nothing left to do. If it isn't executed in a certain amount of time, the current test is marked as failed. In the preceding code snippet, the first thing we did was send a **HyperText Transfer Protocol (HTTP)** GET request to `http://localhost:8000`, which is accepted by our backend server. The expected answer will be in the form of server-side rendered **HyperText Markup Language** (HTML) created through React.

To prove that the response holds this information, we make some assertions in our preceding test, as follows:

1. We use the `should` function from Chai. The great thing is that it's chainable and represents a sentence that directly explains the meaning of what we're doing. The `should.not.exist` function chain makes sure that the given value is empty. The result is `true` if the value is `undefined` or `null`, for example. The consequence is that when the `err` variable is filled, the assertion fails and so our test, `'renders and serves the index page'`, fails too.

2. The same goes for the `should.exist` line. It makes sure that the `res` variable, which is the response given by the backend, is filled. Otherwise, there's a problem with the backend.

3. The `expect` function can also represent a sentence, as with both functions before. We expect `res.statusCode` to have a value of `200`. This assertion can be written as `expect(res.statusCode).to.be.equal(200)`. We can be sure that everything has gone well if the HTTP status is `200`.

4. If nothing has failed so far, we check whether the returned `body` variable, which is the third callback parameter of the `request` function, is valid. For our test scenario, we only need to check whether it contains an `html` tag.

5. We execute the `done` function. We pass the `err` object as a parameter. The result of this function is much like the `should.not.exist` function. If you pass a filled error object to the `done` function, the test fails. The tests become more readable when using the Chai syntax.

If you execute `npm run test` now, you'll receive the following error:

```
Graphbook application test
  1) renders and serves the index page


0 passing (27ms)
1 failing

1) Graphbook application test
     renders and serves the index page:
   Uncaught AssertionError: expected [Error: connect ECONNREFUSED 127.0.0.1:8000] to not exist
     at Object.exist (node_modules/chai/lib/chai/interface/should.js:208:38)
     at Request._callback (test/app.test.js:43:24)
     at self.callback (node_modules/request/request.js:185:22)
     at Request.onRequestError (node_modules/request/request.js:877:8)
     at Socket.socketErrorListener (_http_client.js:406:9)
     at emitErrorNT (internal/streams/destroy.js:92:8)
     at emitErrorAndCloseNT (internal/streams/destroy.js:60:3)
     at processTicksAndRejections (internal/process/task_queues.js:80:21)
```

Figure 11.1 – Failed test because no server ran

Our first `should.not.exist` assertion failed and threw an error. This is because we didn't start the backend when we ran the test. Start the backend in a second terminal with the correct environment variables using `npm run server` and rerun the test. Now, the test is successful, as we can see here:

```
Graphbook application test
  ✓ renders and serves the index page (105ms)

1 passing (110ms)
```

Figure 11.2 – Test passes if the server runs

The output is good, but the process isn't very intuitive. The current workflow is hard to implement when running the tests automatically while deploying your application or pushing new commits to your **version control system** (**VCS**). We'll change this behavior next.

Starting the backend with Mocha

When we want to run a test, the server should start automatically. There are two options to implement this behavior, as outlined here:

- We add the `npm run server` command to the `test` script inside our `package.json` file.

- We import all the necessary files to launch the server within our `app.test.js` file. This allows us to run further assertions or commands against the backend.

The best option is to start the server within our test and not rely on a second command because we can run further tests on the backend. We need to import a further package to allow the server to start within our test, as follows:

```
require('babel-plugin-require-context-hook/register')();
```

We use and execute this package because we load the Sequelize models using the `require.context` function. By loading the package, the `require.context` function is executable for the server-side code. Before we started the server within the test, the plugin hadn't been used, although it was loaded in the `babel-hooks.js` file.

Now, we can load the server directly in the test. Add the following lines of code at the top of the `describe` function, just before the test we've just written:

```
var app;
this.timeout(50000);

before(function(done) {
  app = require('../src/server').default;
  app.on("listening", function() {
    done();
  });
});
```

The idea is to load the server's `index.js` file inside of our test, which starts the backend automatically. To do this, we define an empty variable called `app`. Then, we use `this.timeout` to set the timeout for all tests inside Mocha to `50000` because starting our server, including Apollo Server, takes some time. Otherwise, the test will probably fail because the start time is too long for the standard Mocha timeout.

We must make sure that the server has been completely started before any of our tests are executed. This logic can be achieved with Mocha's `before` function. Using this function, you can set up and configure things such as starting a backend in our scenario. To continue and process all the tests, we need to execute the `done` function to complete the callback of the `before` function. To be sure that the server has started, we do not just run the `done` function after loading the `index.js` file. We bind the `listening` event of the server using the `app.on` function. If the server emits the `listening` event, we can securely run the `done` function, and all tests can send requests to the server. We could also save the return value of the `require` function directly into the `app` variable to hold the `server` object. The problem with this order, however, is that the server may start listening before we can bind the `listening` event. The way we are doing it now makes sure the server hasn't yet started.

The test, however, still isn't working. You'll see an error message that says `'TypeError: app.on is not a function'`. Take a closer look at the server's `index.js` file. At the end of the file, we aren't exporting the `server` object because we only used it to start the backend. This means that the `app` variable in our test is empty and we can't run the `app.on` function. The solution is to export the `server` object at the end of the server's `index.js` file, like this:

```
export default server;
```

You can now execute the test again. Everything should look fine, and all tests should pass.

There is, however, one last problem. If you compare the behavior from the test before importing the server directly in our test or starting it in a second terminal, you might notice that the test isn't finished, or at least the process isn't stopped. Previously, all steps were executed, we returned to the normal shell, and we could execute the next command. The reason for this is that the server is still running in our `app.test.js` file. Therefore, we must stop the backend after all tests have been executed. Insert the following code after the `before` function:

```
after(function(done) {
    app.close(done);
});
```

The `after` function is run when all tests are finished. Our `app` object offers the `close` function, which terminates the server. As a callback, we hand over the `done` function, which is executed once the server has stopped. This means that our test has also finished.

Verifying the correct routing

We now want to check whether all the features of our application are working as expected. One major feature of our application is that React Router redirects the user in the following two cases:

- The user visits a route that cannot be matched.

- The user visits a route that can be matched, but they aren't allowed to view the page.

In both cases, the user should be redirected to the login form. In the first case, we can follow the same approach as for our first test. We send a request to a path that isn't inside our router. Add the following code to the bottom of the `describe` function:

```
describe('404', function() {
    it('redirects the user when not matching path is found',
```

```
      function(done) {
      request({
        url: 'http://localhost:8000/path/to/404',
      }, function(err, res, body) {
        should.not.exist(err);
        should.exist(res);
        expect(res.statusCode).to.be.equal(200);
        assert.ok(res.req.path === '/');
        assert.ok(body.indexOf('<html') !== -1);
        assert.ok(body.indexOf('class="authModal"') !== -1);
        done(err);
      });
    });
  });
});
```

Let's quickly go through all the steps of the preceding test, as follows:

1. We add a new `describe` function to structure our test's output.
2. We send a request inside another `it` function to an unmatched path.
3. The checks are the same as the ones we used when starting the server.
4. We verify that the response's path is the / root. That happens when the redirect is executed. Therefore, we use the `res.req.path === '/'` condition.
5. We check whether the returned `body` variable includes an HTML tag with the `authModal` class. This should happen when the user isn't logged in, and the login or register form is rendered.

If the assertions are successful, we know that React Router works correctly in the first scenario. The second scenario relates to private routes that can only be accessed by authenticated users. We can copy the preceding check and replace the request. The assertions we are doing stay the same, but the **Uniform Resource Locator** (**URL**) of the request is different. Add the following test under the previous one:

```
describe('authentication', function() {
  it('redirects the user when not logged in',
    function(done) {
    request({
```

```
        url: 'http://localhost:8000/app',
    }, function(err, res, body) {
        should.not.exist(err);
        should.exist(res);
        expect(res.statusCode).to.be.equal(200);
        assert.ok(res.req.path === '/');
        assert.ok(body.indexOf('<html') !== -1);
        assert.ok(body.indexOf('class="authModal"') !== -1);
        done(err);
    });
  });
});
```

If an unauthenticated user requests the /app route, they're redirected to the / root path. The assertions verify whether the login form is displayed as before. To differentiate the tests, we add a new describe function so that it has a better structure.

In this section, we have learned how use Mocha to assert that our application works correctly. We are now verifying whether our application starts and also whether the routing works as expected and returns the correct pages.

Next, we want to test the GraphQL API that we built, not only the **server-side rendering (SSR)** functionality.

Testing the GraphQL API with Mocha

We must verify that all the API functions we're offering work correctly. I'm going to show you how to do this with two examples, as follows:

- The user needs to sign up or log in. This is a critical feature where we should verify that the API works correctly.

- The user queries or mutates data via the GraphQL API. For our test case, we will request all chats the logged-in user is related to.

Those two examples should explain all the essential techniques to test every part of your API. You can add more functions that you want to test at any point.

Testing the authentication

We extend the authentication tests of our test with the signup functionality. We're going to send a simple GraphQL request to our backend, including all the required data to sign up a new user. We've already sent requests, so there's nothing new here. In comparison to all the requests before, however, we have to send a POST request, not a GET request. Also, the endpoint for the signup is the /graphql path, where our Apollo Server listens for incoming mutations or queries. Normally, when a user signs up for Graphbook, the authentication token is returned directly, and the user is logged in. We must preserve this token to make future GraphQL requests. We don't use Apollo Client for our test as we don't need to test the GraphQL API.

Create a global variable next to the app variable, where we can store the **JavaScript Object Notation (JSON) Web Token (JWT)** returned after signup, as follows:

```
var authToken;
```

Inside the test, we can set the returned JWT. Add the following code to the authentication function:

```
it('allows the user to sign up', function(done) {
  const json = {
    operationName: null,
    query: "mutation signup($username: String!, $email :
      String!,
    $password : String!) { signup(username: $username,
      email: $email,
    password : $password) { token }}",
    variables: {
      "email": "mocha@test.com",
      "username": "mochatest",
      "password": "123456789"
    }
  };

  request.post({
    url: 'http://localhost:8000/graphql',
    json: json,
```

```
  }, function(err, res, body) {
    should.not.exist(err);
    should.exist(res);
    expect(res.statusCode).to.be.equal(200);
    body.should.be.an('object');
    body.should.have.property('data');
    authToken = body.data.signup.token;
    done(err);
  });
});
```

We begin by creating a `json` variable. This object is sent as a JSON body to our GraphQL API. The content of it should be familiar to you—it's nearly the same format we used when testing the GraphQL API in Postman.

> **Note**
> The JSON we send represents a manual way of sending GraphQL requests.
> There are libraries that you can easily use to save this and directly send the
> query without wrapping it inside an object, such as `graphql-request`:
> `https://github.com/prisma-labs/graphql-request`.

The `json` object includes fake signup variables to create a user with the `mochatest` username. We'll send HTTP `POST` with the `request.post` function. To use the `json` variable, we pass it into the `json` field. The `request.post` function automatically adds the body as a JSON string and the correct `Content-Type` header for you. When the response arrives, we run the standard checks, such as checking for an error or checking an HTTP status code. We also check the format of the returned `body` variable because the response's body variable won't return HTML but will return JSON instead. We make sure that it's an object with the `should.be.an('object')` function. The `should` assertion can directly be used and chained to the `body` variable. If `body` is an object, we check whether there's a `data` property inside. That's enough security to read the token from the `body.data.signup.token` property.

The user is now created in our database. We can use this token for further requests. Be aware that running this test a second time on your local machine is likely to result in a failure because the user already exists. In this case, you can delete it manually from your database. This problem won't occur when running this test while using **continuous integration (CI)**. We'll focus on this topic in the last chapter. Next, we'll make an authenticated query to our Apollo Server and test the result of it.

Testing authenticated requests

We set the authToken variable after the signup request. You could also do this with a login request if a user already exists while testing. Only the query and the assertions we are using are going to change. Also, insert the following code into the before authentication function:

```
it('allows the user to query all chats', function(done) {
  const json = {
    operationName: null,
    query: "query {chats {id users {id avatar username}}}",
    variables: {}
  };

  request.post({
    url: 'http://localhost:8000/graphql',
    headers: {
      'Authorization': authToken
    },
    json: json,
  }, function(err, res, body) {
    should.not.exist(err);
    should.exist(res);
    expect(res.statusCode).to.be.equal(200);
    body.should.be.an('object');
    body.should.have.property('data');
    body.data.should.have.property(
      'chats').with.lengthOf(0);
    done(err);
  });
});
```

As you can see in the preceding code snippet, the json object doesn't include any variables because we only query the chats of the logged-in user. We changed the query string accordingly. Compared to the login or signup request, the chat query requires the user to be authenticated. The authToken variable we saved is sent inside the Authorization header. We now verify again whether the request was successful and check for a data property in the body variable. Notice that, before running the done function, we verify that the data object has a field called chats. We also check the length of the chats field, which proves that it's an array. The length can be statically set to 0 because the user who's sending the query just signed up and doesn't have any chats yet. The output from Mocha looks like this:

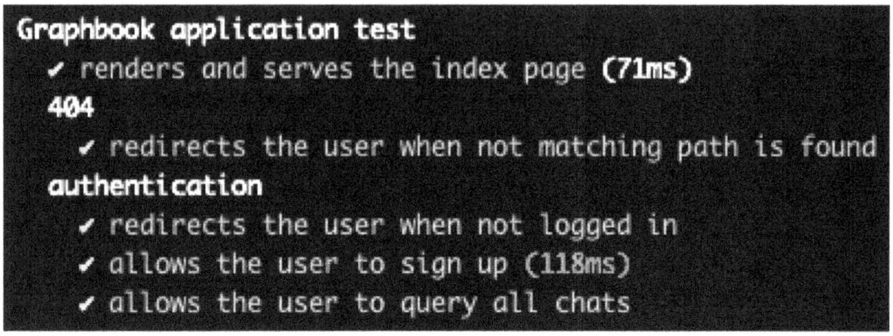

Figure 11. 3 – Authentication test

This is all you need to know to test all the features of your API.

Next, we are going to have a look on Enzyme, which is a great testing tool that allows you to interact with the React components that we have written and ensure that they are working as expected.

Testing React with Enzyme

So far, we've managed to test our server and all GraphQL API functions. Currently, however, we're still missing the tests for our frontend code. While we render the React code when requesting any server route, such as the /app path, we only have access to the final result and not to each component. We should change this to execute the functions of certain components that aren't testable through the backend. First, install some dependencies before using npm, as follows:

```
npm install --save-dev enzyme @wojtekmaj/enzyme-adapter-react-
17ignore-styles jsdom isomorphic-fetch
```

The various packages are described in more detail here:

- The `enzyme` and `@wojtekmaj/enzyme-adapter-react-17` packages provide React with specific features to render and interact with the React tree. This can be through either a real **Document Object Model (DOM)** or shallow rendering. We are going to use a real DOM in this chapter because it allows us to test all features, while shallow rendering is limited to just the first layer of components. We need to rely on a third-party package for the React adapter because there is no official support at the moment for React 17 from Enzyme.

- The `ignore-styles` package strips out all `import` statements for **Cascading Style Sheets (CSS)** files. This is very helpful since we don't need CSS for our tests.

- The `jsdom` package creates a DOM object for us, which is then used to render the React code into.

- The `isomorphic-fetch` package replaces the `fetch` function that all browsers provide by default. This isn't available in Node.js, so we need a polyfill.

We start by importing the new packages directly under the other `require` statements, as follows:

```
require('isomorphic-fetch');
import React from 'react';
import { configure, mount } from 'enzyme';
import Adapter from @wojtekmaj/enzyme-adapter-react-17';
configure({ adapter: new Adapter() });
import register from 'ignore-styles';
register(['.css', '.sass', '.scss']);
```

To use Enzyme, we import React. Then, we create an adapter for Enzyme that supports React 16. We insert the adapter into Enzyme's `configure` statement. Before starting with the frontend code, we import the `ignore-styles` package to ignore all CSS imports. I've also directly excluded **Syntactically Awesome Style Sheets (SASS)** and SCSS files. The next step is to initialize our DOM object, where all the React code is rendered. Here's the code you'll need:

```
const { JSDOM } = require('jsdom');
const dom = new JSDOM('<!doctype html><html><body></body>
  </html>', { url: 'http://graphbook.test' });
const { window } = dom;
global.window = window;
global.document = window.document;
```

We require the `jsdom` package and initialize it with a small HTML string. We don't take the template file that we're using for the server or client because we just want to render our application to any HTML, so how it looks isn't important. The second parameter is an `options` object. We specify a `url` field, which is the host URL, under which we render the React code. Otherwise, we might get an error when accessing `localStorage`. After initialization, we extract the `window` object and define two global variables that are required to mount a React component to our fake DOM. These two properties behave like the `document` and `window` objects in the browser, but instead of the browser, they are global objects inside our Node.js server.

In general, it isn't a good idea to mix up the Node.js `global` object with the DOM of a browser and render a React application in it. Still, we're merely testing our application and not running it in production in this environment, so while it might not be recommended, it helps our test to be more readable. We'll begin the first frontend test with our login form. The visitor to our page can either directly log in or switch to the signup form. Currently, we don't test this switch functionality in any way. This is a complex example, but you should be able to understand the techniques behind it quickly.

To render our complete React code, we're going to initialize an Apollo Client for our test. Import all the dependencies, as follows:

```
import { ApolloClient, InMemoryCache, from } from '@apollo/
client';
import { createUploadLink } from 'apollo-upload-client';
import App from '../src/server/ssr';
```

We also import the `index.js` component of the server-rendered React code. This component will receive our client, which we'll initialize shortly. Add a new `describe` function for all frontend tests, as follows:

```
describe('frontend', function() {
  it('renders and switches to the login or register form',
  function(done) {
    const httpLink = createUploadLink({
      uri: 'http://localhost:8000/graphql',
      credentials: 'same-origin',
    });
    const client = new ApolloClient({
      link: from([
        httpLink
      ]),
```

```
      cache: new InMemoryCache()
    });
  });
});
```

The preceding code creates a new Apollo Client. The client doesn't implement any logic, such as authentication or WebSockets, because we don't need this to test the switch from the login form to the signup form. It's merely a required property to render our application completely. If you want to test components that are only rendered when being authenticated, you can, of course, implement it easily. Enzyme requires us to pass a real React component, which will be rendered to the DOM. Add the following code directly beneath the `client` variable:

```
class Graphbook extends React.Component {
  render() {
    return(
      <App client={client} context={{}} loggedIn={false}
        location= {"/"}/>
    )
  }
}
```

The preceding code is a small wrapper around the App variable that we imported from the server's `ssr` folder. The `client` property is filled with the new Apollo Client. Follow the given instructions to render and test your React frontend code. The following code goes directly under the `Graphbook` class:

1. We use the `mount` function of Enzyme to render the `Graphbook` class to the DOM, as follows:

    ```
    const wrapper = mount(<Graphbook />);
    ```

2. The `wrapper` variable provides many functions to access or interact with the DOM and the components inside it. We use it to prove that the first render displays the login form. The code is illustrated here:

    ```
    expect(wrapper.html()).to.contain('<a>Want to sign up?
    Click here</a>');
    ```

    The `html` function of the `wrapper` variable returns the complete HTML string that has been rendered by the React code. We check this string with the `contain` function of Chai. If the check is successful, we can continue.

3. Typically, the user clicks on the **Want to sign up?** message and React re-renders the signup form. We need to handle this via the `wrapper` variable. The enzyme comes with that functionality innately, as illustrated here:

```
wrapper.find('LoginRegisterForm').find('a').
simulate('click');
```

The `find` function gives us access to the `LoginRegisterForm` component. Inside the markup of the component, we search for an `a` tag, of which there can only be one. If the `find` method returns multiple results, we can't trigger things such as a click because the `simulate` function is fixed to only one possible target. After running both `find` functions, we execute Enzyme's `simulate` function. The only parameter needed is the event that we want to trigger. In our scenario, we trigger a `click` event on the `a` tag, which lets React handle all the rest.

4. We check whether the form was changed correctly by executing the following code:

```
expect(wrapper.html()).to.contain('<a>Want to login?
  Click here</a>');
done();
```

We use the `html` and `contain` functions to verify that everything was rendered correctly. The `done` method of Mocha is used to finish the test.

> **Note**
>
> For a more detailed overview of the API and all the functions that Enzyme provides, have a look at the official documentation: `https://enzymejs.github.io/enzyme/docs/api/`.

This was the easy part. How does this work when we want to verify whether the client can send queries or mutations with authentication? It's actually not that different. We already registered a new user and got a JWT in return. All we need to do is attach the JWT to our Apollo Client, and the router needs to receive the correct `loggedIn` property. The final code for this test looks like this:

```
it('renders the current user in the top bar', function(done) {
  const AuthLink = (operation, next) => {
    operation.setContext(context => ({
      ...context,
      headers: {
        ...context.headers,
        Authorization: authToken
```

```
      },
    }));
    return next(operation);
  };

  const httpLink = createUploadLink({
    uri: 'http://localhost:8000/graphql',
    credentials: 'same-origin',
  });

  const client = new ApolloClient({
    link: from([
      AuthLink,
      httpLink
    ]),
    cache: new InMemoryCache()
  });

  class Graphbook extends React.Component {
    render() {
      return(
        <App client={client} context={{}} loggedIn={true}
          location= {"/app"}/>
      )
    }
  }

  const wrapper = mount(<Graphbook />);
  setTimeout(function() {
    expect(wrapper.html()).to.contain(
    '<div class="user"><img>
    <span>mochatest</span></div>');
    done();
  },2000);
});
```

Here, we are using the `AuthLink` function that we used in the original frontend code. We pass the `authToken` variable to every request that's made by the Apollo Client. In the `Apollo.from` method, we add it before `httpLink`. In the `Graphbook` class, we set `loggedIn` to `true` and `location` to `/app` to render the newsfeed. Because the requests are asynchronous by default and the `mount` method doesn't wait for the Apollo Client to fetch all queries, we couldn't directly check the DOM for the correct content. Instead, we wrapped the assertions and the `done` function in a `setTimeout` function. A timeout of 2,000 **milliseconds** (**ms**) should be enough for all requests to finish and React to have rendered everything. If this isn't enough time, you can increase the number. When all assertions are successful, we can be sure that the `currentUser` query has been run and the top bar has been rendered to show the logged-in user. With these two examples, you should now be able to run any test you want with your application's frontend code.

Summary

In this chapter, we learned all the essential techniques to test your application automatically, including testing the server, the GraphQL API, and the user's frontend. You can apply the Mocha and Chai patterns you learned to other projects to reach a high software quality at any time. Your personal testing time will be greatly reduced.

In the next chapter, we'll have a look at how to improve performance and error logging so that we're always providing a good **user experience** (**UX**).

Section 3: Preparing for Deployment

We are at the end of our journey; you have come to the point where you have finished the main part of your work. The last part that is left is to actually release your application to a big audience. In this section, you will learn how to deploy your application safely and continuously to AWS through CircleCI.

In this section, there are the following chapters:

- *Chapter 12, Continuous Deployment with CircleCI and AWS*

12
Continuous Deployment with CircleCI and AWS

In the last two chapters, we prepared our application through tests with Mocha. We have built an application that is ready for the production environment.

We will now generate a production build that's ready for deployment. We've arrived at the point where we can set up **Amazon Elastic Container Service (Amazon ECS)** and implement the ability to build and deploy Docker images through a continuous deployment workflow.

The process of continuous deployment will help to keep changes small for the production environment. Keeping changes in your application continuous and small will make issues trackable and fixable, whereas releasing a set of multiple features at once will leave the location for bugs open for investigation as multiple things will have changed with just one release.

This chapter covers the following topics:

- Production-ready bundling
- What is Docker?
- Configuring Docker
- Setting up AWS RDS (short for **AWS Relational Database Service**) as a production database
- What is continuous integration/deployment?
- Setting up continuous deployment with CircleCI
- Deploying our application to **Amazon Elastic Container Registry** (**Amazon ELB**) and ECS with AWS **Application Load Balancer** (**ALB**)

Technical requirements

The source code for this chapter is available in the following GitHub repository:

`https://github.com/PacktPublishing/Full-Stack-Web-Development-with-GraphQL-and-React-2nd-Edition/tree/main/Chapter12`

Preparing the final production build

We have come a long way to get here. Now is the time when we should look at how we currently run our application and how we should prepare it for a production environment.

Currently, we use our application in a development environment while working on it. It is not highly optimized for performance or low-bandwidth usage. We include developer functionalities with the code so that we can debug it properly.

For use in a real production environment, we should only include what is necessary for the user. When setting the NODE_ENV variable to production, we remove most of the unnecessary development mechanics.

By bundling our server-side code, we will get rid of unnecessary loading times and will improve the performance. To bundle our backend code, we are going to set up a new webpack configuration file. Follow these instructions:

1. Install the following two dependencies:

    ```
    npm install --save-dev webpack-node-externals @babel/
    plugin-transform-runtime
    ```

These packages do the following:

- The `webpack-node-externals` package gives you the option to exclude specific modules while bundling your application with webpack. It reduces the final bundle size.

- The `@babel/plugin-transform-runtime` package is a small plugin that enables us to reuse Babel's helper methods, which usually get inserted into every processed file. It reduces the final bundle size.

2. Create a `webpack.server.build.config.js` file next to the other webpack files with the following content:

```
const path = require('path');
const nodeExternals = require('webpack-node-externals');
const buildDirectory = 'dist/server';

module.exports = {
  mode: 'production',
  entry: [
    './src/server/index.js'
  ],
  output: {
    path: path.join(__dirname, buildDirectory),
    filename: 'bundle.js',
    publicPath: '/server'
  },
  module: {
    rules: [{
      test: /\.js$/,
      use: {
        loader: 'babel-loader',
        options: {
          plugins: ["@babel/plugin-transform-runtime"]
        }
      },
    }],
  },
  node: {
```

```
        __dirname: false,
        __filename: false,
    },
    target: 'node',
    externals: [nodeExternals()],
    plugins: [],
};
```

The preceding configuration file is very simple and not complex. Let's go through the settings that we use to configure webpack:

- We load our new `webpack-node-externals` package at the top.
- The `build` directory, where we save the bundle, is in the `dist` folder, inside of a special `server` folder.
- The `mode` field is set to `'production'`.
- The `entry` point for webpack is the server's root `index.js` file.
- The `output` property holds the standard fields to bundle our code and saves it inside of the folder specified through the `buildDirectory` variable.
- We use the previously installed `@babel/plugin-transform-runtime` plugin in the `module` property to reduce the file size for our bundle.
- Inside of the `node` property, you can set Node.js-specific configuration options. The `__dirname` field tells webpack that the global `__dirname` variable is used with its default settings and is not customized by webpack. The same goes for the `__filename` property.
- The `target` field accepts multiple environments in which the generated bundle should work. For our case, we set it to `'node'`, as we want to run our backend in Node.js.
- The `externals` property gives us the possibility to exclude specific dependencies from our bundle. By using the `webpack-node-externals` package, we prevent all `node_modules` packages from being included in our bundle.

3. To make use of our new build configuration file, we add two new commands to the `scripts` field of our `package.json` file. As we are trying to generate a final production build that we can publicize, we have to build our client-side code in parallel. Add the following two lines to the `scripts` field of the `package.json` file:

```
"build": "npm run client:build && npm run server:build",
"server:build": "webpack --config webpack.server.build.
config.js"
```

The `build` command uses the `&&` syntax to chain two `npm run` commands. It executes the build process for our client-side code first, and afterward, it bundles the entire server-side code. The result is that we have a filled `dist` folder with a `client` folder and a `server` folder. Both can import components dynamically.

4. To start our server with the new production code, we are going to add one further command to the `scripts` field. The old `npm run server` command would start the server-side code in the unbundled version, which is not what we want. Insert the following line into the `package.json` file:

```
"server:production": "node dist/server/bundle.js"
```

The preceding command simply executes the `bundle.js` file from the `dist/server` folder, using the plain `node` command to launch our backend.

Now, you should be able to generate your final build by running `npm run build`. Before starting the production server as a test, however, make sure that you have set all environment variables for your database correctly, or your `JWT_SECRET`, for example. Then, you can execute the `npm run server:production` command to launch the backend.

5. Our tests need to be run in a way to reflect the same production conditions, because only then can we verify that all features that are enabled in the live environment work correctly. To make sure that is true, we need to change how we execute the tests. Edit the `test` command of the `package.json` file to reflect this change, as follows:

```
"test": "npm run build && mocha --exit test/ --require babel-hook --require @babel/polyfill --recursive",
```

Now, you should be able to test your application with the same generated production bundles.

In the next section, we will cover how to use Docker to bundle your entire application.

Setting up Docker

Publishing an application is a critical step that requires a lot of work and care. Many things can go wrong when releasing a new version. We have already made sure that we can test our application before it goes live.

The real act of transforming our local files into a production-ready package, which is then uploaded to a server, is the most onerous task. Regular applications generally rely on a server that is preconfigured with all the packages that the application needs to run. For example, when looking at a standard PHP setup, most people rent a preconfigured server. This means that the PHP runtime, with all the extensions, such as the MySQL PHP library, are installed via the built-in package manager of the operating system. This procedure applies not only to PHP but also to nearly any other programming language. This might be okay for general websites or applications that are not too complex, but for professional software development or deployment, this process can lead to issues, such as the following:

- The configuration needs to be done by someone that knows the requirements of the application and the server itself.

- A second server needs the same configuration in order to allow our application to run. While doing that configuration, we must ensure that all servers are standardized and consistent with one another.

- All of the servers have to be reconfigured when the runtime environment gets an update, either because the application requires it, or due to other reasons, such as security updates. In this case, everything must be tested again.

- Multiple applications running inside of the same server environment may require different package versions or may interfere with each other.

- The deployment process must be executed by someone with the required knowledge.

- Starting an application directly on a server exposes it to all the services running on the server. Other processes could take over your complete application since they run within the same environment.

- Also, the application is not limited to using a specified maximum of the server's resources.

Many people have tried to figure out how to avoid these consequences by introducing a new containerization and deployment workflow.

What is Docker?

One of the most important pieces of software is called Docker. It was released in 2013, and its function is to isolate the application within a container by offering its own runtime environment, without having access to the server itself.

The aim of a container is to isolate the application from the operating system of the server.

Standard virtual machines can also accomplish this by running a guest operating system for the application. Inside of the virtual machine, all packages and runtimes can be installed to prepare it for your application. This solution comes with significant overhead, of course, because we are running a second operating system that's just for our application. It is not scalable when many services or multiple applications are involved.

On the other hand, Docker containers work entirely differently. The application itself, and all of its dependencies, receive a segment of the operating system's resources. All processes are isolated by the host system inside of those resources.

Any server supporting the container runtime environment (which is Docker) can run your dockerized application. The great thing is that the actual operating system is abstracted away. Your operating system will be very slim, as nothing more than the kernel and Docker is required.

With Docker, the developer can specify how the container image is composed. They can directly test and deploy those images on their infrastructure.

To see the process and advantages that Docker provides, we are going to build a container image that includes our application and all the dependencies it needs to run.

Installing Docker

As with any virtualization software, Docker has to be installed via the regular package manager of your operating system.

I will assume that you are using a Debian-based system. If this is not the case, please get the correct instructions for your system at `https://docs.docker.com/install/overview/`.

Continue with the following instructions to get Docker up and running:

1. Update your system's package manager, as follows:

    ```
    sudo apt-get update
    ```

2. We can install the Docker package to our system, as follows:

    ```
    sudo apt-get install docker
    ```

    If you are running an Ubuntu version with snap installed, you can also use the following command:

    ```
    sudo snap install docker
    ```

That's everything that is required to get a working copy of Docker on your system.

Next, you will learn how to use Docker by building your first Docker container image.

Dockerizing your application

Many companies have adopted Docker and replaced their old infrastructure setup, thereby largely reducing system administration. Still, there is some work to do before deploying your application straight to production.

One primary task is to dockerize your application. The term **dockerize** means that you take care of wrapping your application inside of a valid Docker container.

There are many service providers that connect Docker with CI or continuous deployment because they work well together. In the last section of this chapter, you will learn what continuous deployment is and how it can be implemented. We are going to rely on such a service provider. It will provide us with an automatic workflow for our continuous deployment process. Let's first start dockerizing our application.

Writing your first Dockerfile

The conventional approach to generating a Docker image of your application is to create a `Dockerfile` at the root of your project. But what is a `Dockerfile` for?

A `Dockerfile` is a series of commands that are run through the Docker **Command-Line Interface (CLI)**. The typical workflow in such a file looks as follows:

1. A `Dockerfile` starts from a base image, which is imported using the `FROM` command. This base image may include a runtime environment, like Node.js, or other things that your project can make use of. The container images are downloaded from Docker Hub, which is a central container registry that you can find at `https://hub.docker.com/`. There is the option to download the images from custom registries, too.

2. Docker offers many commands to interact with the image and your application code. Those commands can be looked up at `https://docs.docker.com/engine/reference/builder/`.

3. After the configuration of the image has finished and all build steps are complete, you will need to provide a command that will be executed when your application's Docker container starts.

4. The result of the build steps will be a new Docker image (see *Figure 12.1*). The image is saved on the machine where it was generated.

5. Optionally, you can now publish your new image to a registry, where other applications or users can pull your image. You can also upload them as private images or private registries.

We will start by generating a simple Docker image. First, create the Dockerfile inside of the root of your project. The filename is written without any file extensions.

The first task is to find a matching base image that we can use for our project. The criteria by which we choose a base image are the dependencies and runtime environment. As we have mainly used Node.js without relying on any other server-side package that needs to be covered from our Docker container, we only need to find a base image that provides Node.js. For the moment, we will ignore the database, and we'll focus on it again in a later step.

Docker Hub is the official container image registry, providing many minimalistic images. Just insert the following line inside of our new Dockerfile, in the root of our project:

```
FROM node:14
```

As we mentioned before, we use the FROM command to download our base image. As the name of the preceding image states, it includes Node.js in version 14. There are numerous other versions that you can use. Beyond the different versions, you can also find different flavors (for example, a Node.js image based on an Alpine Linux image). Take a look at the image's README to get an overview of the available options at https://hub.docker.com/_/node/.

> **Important Note**
>
> I recommend that you read through the reference documentation of the Dockerfile. Many advanced commands and scenarios are explained there, which will help you to customize your Docker workflow. Just go to https://docs.docker.com/engine/reference/builder/.

After Docker has run the FROM command, you will be working directly within this base image, and all further commands will then run inside of this environment. You can access all features that the underlying operating system provides. Of course, the features are limited by the image that you have chosen. A Dockerfile is only valid if it starts with the FROM command.

The next step for our Dockerfile is to create a new folder, in which the application will be stored and run. Add the following line to the Dockerfile:

```
WORKDIR /usr/src/app
```

The WORKDIR command changes the directory to the specified path. The path that you enter lives inside of the filesystem of the image, which does not affect your computer's filesystem. From then on, the RUN, CMD, ENTRYPOINT, COPY, and ADD Docker commands will be executed in the new working directory. Furthermore, the WORKDIR command will create a new folder if it does not exist yet.

Next, we need to get our application's code inside of the new folder. Until now, we have only made sure that the base image was loaded. The image that we are generating does not include our application yet. Docker provides a command to move our code into the final image.

As the third line of our Dockerfile, add the following code:

```
COPY . .
```

The COPY command accepts two parameters. The first one is the source, which can be a file or folder. The second parameter is the destination path inside of the image's filesystem. You can use a subset of regular expressions to filter the files or folders that you copy.

After Docker has executed the preceding command, all contents living in the current directory will be copied over to the /usr/src/app path. The current directory, in this case, is the root of our project folder. All files are now automatically inside of the final Docker image. You can interact with the files through all Docker commands but also with the commands that the shell provides.

One important task is that we install all npm packages that our application relies on. When running the COPY command, such as in the preceding code, all files and folders are transferred, including the node_modules folder. This could lead to problems when trying to run the application, however. Many npm packages are compiled when they are being installed, or they differentiate between operating systems. We must make sure that the packages that we use are clean, and work in the environment that we want them to work in. We must do two things to accomplish this, as follows:

1. Create a .dockerignore file in the root of the project folder, next to the Dockerfile, and enter the following content:

    ```
    node_modules
    ```

    The .dockerignore file is comparable to the .gitignore file, which excludes special files or folders from being tracked by Git. Docker reads the .dockerignore file before all files are sent to the Docker daemon. If it is able to read a valid .dockerignore file, all specified files and folders are excluded. The preceding two lines exclude the whole node_modules folder.

2. Install the npm packages inside of Docker. Add the following line of code to the `Dockerfile`:

```
RUN npm install
```

The RUN command executes npm `install` inside of the current working directory. The related `package.json` file and `node_modules` folder are stored in the filesystem of the Docker image. Those files are directly committed and are included in the final image. Docker's RUN command sends the command that we pass as the first parameter into Bash and executes it. To avoid the problems of spaces in the shell commands, or other syntax problems, you can pass the command as an array of strings, which will be transformed by Docker into valid Bash syntax. Through RUN, you can interact with other system-level tools (such as `apt-get` or `curl`).

Now that all files and dependencies are in the correct filesystem, we can start Graphbook from our new Docker image. Before doing so, there are two things that we need to do – we have to allow for external access to the container via the IP and define what the container should do when it has started.

3. Graphbook uses port 8000 by default, under which it listens for incoming requests, be it a GraphQL or normal web request. When running a Docker container, it receives its own network, with an IP and ports. We must make port 8000 available to the public, not only inside of the container itself. Insert the following line at the end of the `Dockerfile` to make the port accessible from outside of the container:

```
EXPOSE 8000
```

It is essential that you understand that the EXPOSE command does not map the inner port 8000 from the container to the matching port of our working machine. By writing the EXPOSE command, you give the developer using the image the option to publish port 8000 to any port of the real machine running the container. The mapping is done while starting the container, not when building the image. Later in this chapter, we will look at how to map port 8000 to a port of your local machine.

4. Finally, we have to tell Docker what our container should do once it has booted. In our case, we want to start our backend (including SSR, of course). Since this should be a simple example, we will start the development server.

    Add the last line of the `Dockerfile`, as follows:

```
CMD [ "npm", "run", "server" ]
```

The CMD command defines the way that our container is booted, and which command to run. We are using the exec option of Docker to pass an array of strings. A Dockerfile can only have one CMD command. The exec format does not run a Bash or shell command when using CMD.

The container executes the server script of our package.json file, which has been copied into the Docker image.

At this point, everything is finished and prepared to generate a basic Docker image. Next, we will continue with getting a container up and running.

Building and running Docker containers

The Dockerfile and .dockerignore files are ready. Docker provides us with the tools to generate a real image, which we can run or share with others. Having a Dockerfile on its own does not make an application dockerized.

Make sure that the database credentials specified in the /server/config/index.js file for the backend are valid for development because they are statically saved there. Furthermore, the MySQL host must allow for remote connections from inside the container.

Execute the following command to build the Docker image on your local machine:

```
docker build -t sgrebe/graphbook .
```

This command requires you to have the Docker CLI and daemon installed.

The first option that we use is -t, following a string (in our case, sgrebe/graphbook). The finished build will be saved under the username sgrebe and the application name graphbook. This text is also called a tag. The only required parameter of the docker build command is the build context or the set of files that Docker will use for the container. We specified the current directory as the build context by adding the dot at the end of the command. Furthermore, the build action expects the Dockerfile to be located within this folder. If you want the file to be taken from somewhere else, you can specify it with the --file option.

> **Important Note**
>
> If the docker build command fails, it may be that some environment variables are missing. They usually include the IP and port of the Docker daemon. To look them up, execute the docker-machine env command and set the environment variables as returned by the command.

When the command has finished generating the image, it should be available locally. To prove this, you can use the Docker CLI by running the following command:

```
docker images
```

The output from Docker should look as follows:

REPOSITORY	TAG	IMAGE ID	CREATED	SIZE
sgrebe/graphbook	latest	fe30bceb0268	27 minutes ago	1.22GB
node	10	75a3a4428e1d	3 days ago	894MB

Figure 12.1 – Docker images

You should see two containers; the first one is the `sgrebe/graphbook` container image, or whatever you used as a tag name. The second one should be the `node` image, which we used as the base for our custom Docker image. The size of the custom image should be much higher because we installed all npm packages.

Now, we should be able to start our Docker container with this new image. The following command will launch your Docker container:

```
docker run -p 8000:8000 -d --env-file .env sgrebe/graphbook
```

The `docker run` command also has only one required parameter, which is the image to start the container with. In our case, this is `sgrebe/graphbook`, or whatever you specified as a tag name. Still, we define some optional parameters that we need to get our application working. You can find an explanation of each of them, as follows:

- We set the `-p` option to `8000:8000`. The parameter is used to map ports from the actual host operating system to a specific port inside of the Docker container. The first port is the port of the host machine, and the second one is the port of the container. This option gives us access to the exposed port `8000`, where the application is running under `http://localhost:8000` of our local machine.

- The `--env-file` parameter is required to pass environment variables to the container. Those can be used to hand over the `NODE_ENV` or `JWT_SECRET` variables, for example, which we require throughout our application. We will create this file in a second.

- You can also pass the environment variables one by one using the `-e` option. It is much easier to provide a file, however.

- The `-d` option sets the container to **detached mode**. This means that your container will not run in the foreground after executing it inside the shell. Instead, after running the command, you will have access to the shell again and will see no output from the container. If you remove the option again, you will see all of the logs that our application triggers.

> **Important Note**
>
> The `docker run` command provides many more options. It allows for various advanced setups. The link to the official documentation is `https://docs.docker.com/engine/reference/run/#general-form`.

Let's create the `.env` file in the root directory of our project. Insert the following content, replacing all placeholders with the correct value for every environment variable:

```
NODE_ENV=development
JWT_SECRET=YOUR_JWT_SECRET
AWS_ACCESS_KEY_ID=YOUR_AWS_KEY_ID
AWS_SECRET_ACCESS_KEY=YOUR_AWS_SECRET_ACCESS_KEY
```

The `.env` file is a simple key-value list, where you can specify one variable per line, which our application can access from its environment variables.

It is vital that you do not commit this file to the public at any stage. Please add this file directly to the `.gitignore` file.

If you have filled out this file, you will be able to start the Docker container with the previous command that I showed you. Now that the container is running in detached mode, you will have the problem that you cannot be sure whether Graphbook has started to listen. Consequently, Docker also provides a command to test this, as follows:

```
docker ps
```

The `docker ps` command gives you a list of all running containers. You should find the Graphbook container in there, too. The output should appear as follows:

```
CONTAINER ID   IMAGE             COMMAND             CREATED         STATUS         PORTS                      NAMES
08499322a998   sgrebe/graphbook   "npm run server"    4 seconds ago   Up 3 seconds   0.0.0.0:8000->8000/tcp    dreamy_knuth
```

12.2 – Docker running containers

> **Important Note**
>
> Like all commands that Docker provides, the `docker ps` command gives us many options to customize and filter the output. Read up on all of the features that it offers in the official documentation at `https://docs.docker.com/engine/reference/commandline/ps/`.

Our container is running, and it uses the database that we have specified. You should be able to use Graphbook as you know it by visiting `http://localhost:8000`.

If you take a look at the preceding figure, you will see that all running containers receive their own IDs. This ID can be used in various situations to interact with the container.

In development, it makes sense to have access to the command-line printouts that our application generates. When running the container in detached mode, you have to use the Docker CLI to see the printouts, using the following command. Replace the ID at the end of the command with the ID of your container:

```
docker logs 08499322a998
```

The docker logs command will show you all the printouts that have been made by our application or container recently. Replace the preceding ID with the one given to you by the docker ps command. If you want to see the logs in real time while using Graphbook, you can add the --follow option.

As we are running the container in detached mode, you will not be able to stop it by just using *Ctrl + C* as before. Instead, you have to use the Docker CLI again.

To stop the container again, run the following command:

```
docker stop 08499322a998
```

To finally remove it, run the following command:

```
docker rm 08499322a998
```

The docker rm command stops and removes the container from the system. Any changes made to the filesystem inside of the container will be lost. If you start the image again, a new container will be created, with a clean filesystem.

When working and developing with Docker frequently, you will probably generate many images to test and verify the deployment of your application. These take up a lot of space on your local machine. To remove the images, you can execute the following command:

```
docker rmi fe30bceb0268
```

The ID can be taken from the docker images command, the output of which you can see in the first image in this section. You can only remove an image if it is not used in a running container.

We have come far. We have successfully dockerized our application. However, it is still running in development mode, so there is a lot to do.

Multi-stage Docker production builds

Our current Docker image, which we are creating from the `Dockerfile`, is already useful. We want our application to be transpiled and running in production mode because many things are not optimized for the public when running in development mode.

Obviously, we have to run our build scripts for the backend and frontend while generating the Docker image.

Up until now, we have installed all npm packages and copied all files and folders for our project to the container image. This is fine for development because this image is not published or deployed to a production environment. When going live with your application, you will want your image to be as slim and efficient as possible. To achieve this, we will use a so-called **multi-stage build**.

Before Docker implemented the functionality to allow for multi-stage builds, you had to rely on tricks, such as using shell commands to only keep the files that were really required in the container image. The problem that we have is that we copy all files that are used to build the actual distribution code from the project folder. Those files are not needed in the production Docker container, however.

Let's see how this looks in reality. You can back up or remove the first `Dockerfile` that we wrote, as we will start with a blank one now. The new file still needs to be called `Dockerfile`. All the following lines of code go directly into this empty `Dockerfile`. Follow these instructions to get the multi-stage production build running:

1. Our new file starts with the `FROM` command again. We are going to have multiple `FROM` statements because we are preparing a multi-stage build. The first one should look as follows:

    ```
    FROM node:14 AS build
    ```

    We are introducing the first build stage here. As before, we are using the `node` image in version 14. Furthermore, we append the `AS build` suffix, which tells Docker that this stage, and everything that we do in it, will be accessible under the name `build` later. A new stage is started with every new `FROM` command.

2. Next, we initialize the working directory, as we did in our first `Dockerfile`, as follows:

    ```
    WORKDIR /usr/src/app
    ```

3. It is essential to only copy the files that we really need. It hugely improves the performance if you reduce the number of files that need to be processed:

```
COPY .babelrc ./
COPY package*.json ./
COPY webpack.server.build.config.js ./
COPY webpack.client.build.config.js ./
COPY src src
COPY assets assets
COPY public public
```

We copy the `.babelrc`, `package.json`, `package-lock.json`, and webpack files that are required for our application. These include all information we need to generate a production build for the frontend and backend. Furthermore, we also copy the `src`, `public`, and `assets` folders, because they include the code and CSS that will be transpiled and bundled.

4. Like in our first `Dockerfile`, we must install all npm packages; otherwise, our application won't work. We do this with the following line of code:

```
RUN npm install
```

5. After all the packages have been installed successfully, we can start the build process. We added the `build` script in the first section of this chapter. Add the following line to execute the script that will generate the production bundles in the Docker image:

```
RUN npm run build
```

The following command will generate a `dist` folder for us, where the runnable code (including CSS) will be stored. After the `dist` folder with all bundles has been created, we will no longer need most of the files that we initially copied over to the current build stage.

6. To get a clean Docker image that only contains the `dist` folder and the files that we need to run the application, we will introduce a new build stage that will generate the final image. The new stage is started with a second FROM statement, as follows:

```
FROM node:14
```

We are building the final image in this build step; therefore, it does not need its own name.

7. Again, we need to specify the working directory for the second stage, as the path is
    not copied from the first build stage:

    ```
    WORKDIR /usr/src/app
    ```

8. Before continuing, we need to ensure that the application has access to all
    environment variables. For that, add the following lines to the Dockerfile:

    ```
    ENV NODE_ENV production
    ENV JWT_SECRET JWT_SECRET
    ENV username YOUR_USERNAME
    ENV password YOUR_PASSWORD
    ENV database YOUR_DATABASE
    ENV host YOUR_HOST
    ENV AWS_ACCESS_KEY_ID AWS_ACCESS_KEY_ID
    ENV AWS_SECRET_ACCESS_KEY AWS_SECRET_ACCESS_KEY
    ```

    We use the ENV command from Docker to fill the environment variables while
    building the image.

9. Because we have given our first build stage a name, we can access the filesystem of this
    stage through that name. To copy the files from the first stage, we can add a parameter
    to the COPY statement. Add the following commands to the Dockerfile:

    ```
    COPY --from=build /usr/src/app/package.json package.json
    COPY --from=build /usr/src/app/dist dist
    COPY start.sh start.sh
    COPY src/server src/server
    ```

    As you should see in the preceding code, we are copying the package.json file
    and the dist folder. However, instead of copying the files from our original project
    folder, we are getting those files directly from the first build stage. For this, we use
    the --from option, following the name of the stage that we want to access; so, we
    enter the name build. The package.json file is needed because it includes all
    dependencies and the scripts field, which holds the information on how to run
    the application in production. The dist folder is, of course, our bundled application.

    Furthermore, we copy a start.sh file that we will create and the server folder
    because in there we have all the database migrations.

10. Note that we only copy the package.json file and the dist folder. Our npm dependencies are not included in the application build inside of the dist folder. As a result, we need to install the npm packages in the second build stage, too:

```
RUN npm install --only=production
```

The production image should only hold npm packages that are really required; npm offers the only parameter, which lets you install only the production packages, as an example. It will exclude all devDependecies of your package.json file. This is really great for keeping your image size low.

Then, there are three npm packages that are technically not a dependency, which is defined in our package.json file, because they are not required to get our application running. Still, they are needed to get our database migrations applied. Add the following RUN command to the Dockerfile:

```
RUN npm install -g mysql2 sequelize sequelize-cli
```

The above command will install Sequelize, the Sequelize CLI, and the mysql2 package. We will leverage them to apply migrations at the start of the Docker container. There are also further ways to trigger them manually and not at the start of a Docker container, but this will work for our setup.

11. The last two things to do here are to expose the container port to the public and to execute the CMD command, which will let the image run a command of our package.json file when the container has booted:

```
EXPOSE 8000
CMD [ "sh", "start.sh" ]
```

12. Lastly, we need to create a start.sh file at the root of the project with the following content:

```
sequelize db:migrate --migrations-path src/server/
migrations --config src/server/config/index.js --env
production
npm run server:production
```

In the start.sh file, we have two lines. The first one runs all database migrations. The last one starts the server based on the generated production bundle.

Now, you can execute the `docker build` command again and try to start the container. There is only one problem – the database credentials are read from the environment variables when running in production. As the production setup for a database cannot be on our local machine, it needs to live somewhere on a real server. We could also accomplish this through Docker, but this would involve a very advanced Docker configuration. We would need to save the MySQL data in separate storage because Docker does not persist data of any kind, by default.

Personally, I like to rely on a cloud host, which handles the database setup for me. It is not only great for the overall setup but also improves the scalability of our application. The next section will cover Amazon RDS and how to configure it for use with our application. You can use any database infrastructure that you like.

Amazon RDS

AWS offers Amazon **RDS**, which is an easy tool for setting up a relational database in just a few clicks. Shortly, I will explain how to create your first database with RDS, and afterward, we will look at how to insert environment variables correctly in order to get a database connection going with our application.

The first step is to log in to the AWS Management Console, as we did in *Chapter 7, Handling Image Uploads*. You can find the service by clicking on the **Services** tab in the top bar and searching for RDS.

After navigating to **RDS**, you will see the dashboard, as shown in the following screenshot:

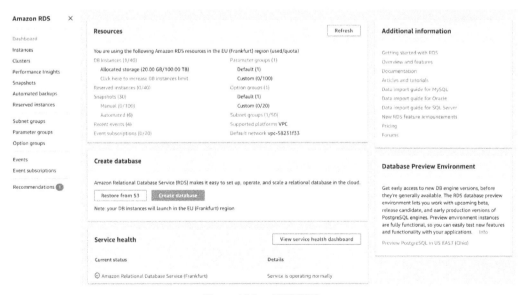

Figure 12.3 – AWS RDS

Follow these instructions to set the RDS database up:

1. Initialize a new database by hitting the **Create database** button. You will be
    presented with a new screen, where you should select an engine for our new
    database and how to create it, as shown in the following screenshot:

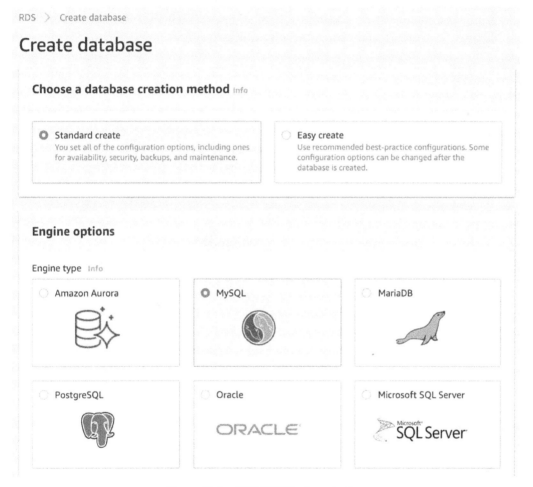

Figure 12.4 – AWS RDS Engine selection

I recommend that you select **MySQL** here. You should also be able to select **Amazon Aurora** or **MariaDB**, as they are also MySQL compatible; for this book, I have chosen MySQL. Also, stay with the standard creation method. Continue by scrolling down.

2. You will need to specify the use case for your database. The production option is only recommended for live applications because this will include higher costs. Choose the free tier, as shown in the following screenshot:

Figure 12.5 – AWS RDS Templates selection

3. Continue by scrolling down. Next, you need to fill in the database credentials to authenticate on the backend later. Fill in the details, as shown here:

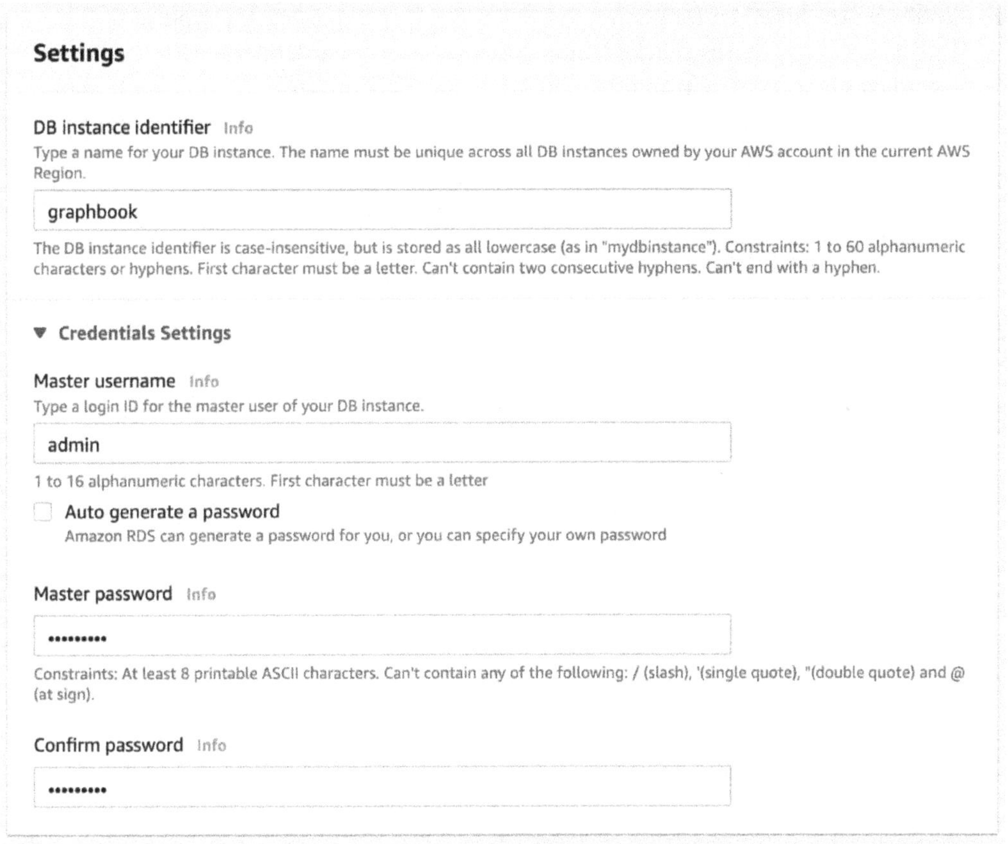

Figure 12.6 – AWS RDS database credentials

4. Next, you need to select the AWS instance and, by that, the computing power of the database. Select **db.t2.micro**, which is free and enough for our use case now. It should look like as follows:

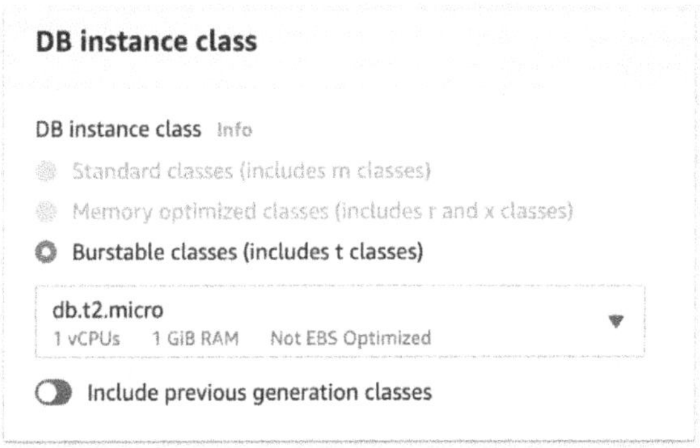

Figure 12.7 – AWS RDS instance class

5. You will now be asked for connectivity settings. It is important that you select **Public access**, with **Yes** checked. This does not share your database with the public but makes it accessible from other IPs and other EC2 instances if you select them in your AWS security group. Also, you need to create a new subnet group and give a new security group name:

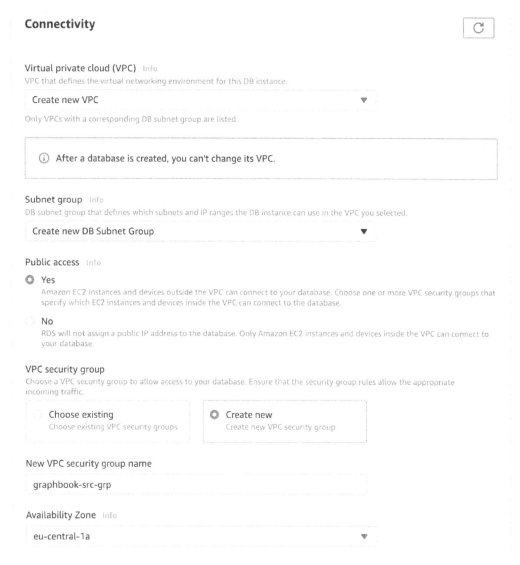

Figure 12.8 – AWS RDS network settings

6. In the **Additional configuration** section, we need to provide an initial database name, which will create a database inside RDS after finishing the setup:

Figure 12.9 – The Additional configuration window

7. Finish the setup process for your first AWS RDS database by clicking on **Create database** at the bottom of the screen.

You should now be redirected to the list of all databases.

Click on the new database instance that has been created. If you scroll down, you will see a list of security groups. Click on the group with the **CIDR/IP - Inbound** type:

Security group		Type		Rule	
graphbook-src-grp (sg-0a08126acbd582acb)		CIDR/IP - Inbound		92.97.27.239/32	
graphbook-src-grp (sg-0a08126acbd582acb)		CIDR/IP - Outbound		0.0.0.0/0	

Figure 12.10 – AWS security group rules

If you click on the first rule, you will be able to insert the IP that is allowed to access the database. If you insert the 0.0.0.0 IP, it will allow any remote IP to access the database. This is not a recommended database setup for production use, but it makes it easier to test it with multiple environments in developmental use.

The credentials that you have specified for the database must be included in the `.env` file for running our Docker container, as follows:

```
username=YOUR_USERNAME
password=YOUR_PASSWORD
database=YOUR_DATABASE
host=YOUR_HOST
```

The `host` URL can be taken from the Amazon RDS instance dashboard. It should look something like `INSTANCE_NAME.xxxxxxxxx.eu-central-1.rds.amazonaws.com`.

Now, you should be able to run the build for your Docker image again, without any problems. The database has been set up and is available.

Next, we will look at how we can automate the process of generating the Docker image through continuous integration.

Configuring continuous integration

Many people (especially developers) will have heard of **continuous integration** (**CI**) or **continuous deployment** (**CD**). However, most of them cannot explain their meanings and the differences between the two terms. So, what is CI and CD, in reality?

When it comes to releasing your application, it might seem easy to upload some files to a server and then start the application through a simple command in the shell, via SSH.

This approach might be a solution for many developers or small applications that are not updated often. However, for most scenarios, it is not a good approach. The word *continuous* represents the fact that all changes or updates are continuously either tested, integrated, or even released. This would be a lot of work, and it would be tough to do if we stayed with a simple file upload and took a manual approach. Automating this workflow makes it convenient to update your application at any time.

CI is the development practice where all developers commit their code to the central project repository at least once a day to bring their changes to the mainline stream of code. The integrated code will be verified by automated test cases. This will avoid problems when trying to go live at a specific time.

CD goes further; it's based on the main principles of CI. Every time the application is successfully built and tested, the changes are directly released to the customer. This is what we are going to implement.

Our automation process will be based on **CircleCI**. It is a third-party service offering a CI and CD platform, with a massive number of features.

To sign up for CircleCI, visit `https://circleci.com/signup/`.

You will need to have a Bitbucket or GitHub account in order to sign up. This will also be the source from which the repositories of your application will be taken, for which we can begin using CI or CD.

To get your project up and running with CircleCI, you will need to click on the **Projects** button in the left-hand panel, or you will be redirected there because you have no projects set up yet. After signing up, you should see all your repositories inside of CircleCI.

Select the project that you want to process with CircleCI by hitting **Set Up Project** on the right-hand side of the project. You will then be confronted with the following screen:

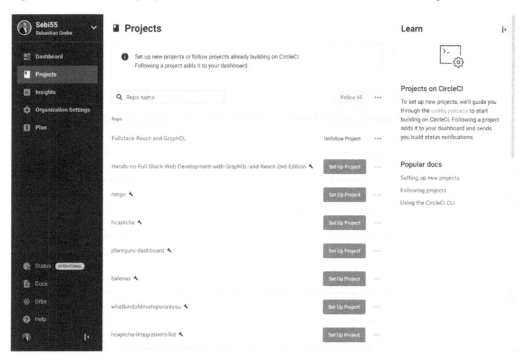

Figure 12.11 – CircleCI Projects

The problem is that you have not configured your repository or application accordingly. You are required to create a folder called `.circleci` and a file inside of it, called `config.yml`, which tells CircleCI what to do when a new commit is pushed to the repository. CircleCI will either ask you to set it up yourself or it will do it for you. I recommend selecting that we are going to do it on our own, as this book guides you through the steps.

Next, set **sample config** as **Node**. The final step will be to push a new commit with the matching CircleCI config.

We will create a straightforward CircleCI configuration so that we can test that everything is working. The final configuration will be done at a later step when we have configured Amazon ECS, which will be the host of our application.

So, create a `.circleci` folder at the root of our project and a `config.yml` file inside of this new folder. The `.yml` file extension stands for **YAML**, which is a file format for saving various configurations or data. What is important here is that all `.yml` files need a correct indentation. Otherwise, they will not be valid files and cannot be understood by CircleCI.

Insert the following code into the `config.yml` file:

```
version: 2.1
jobs:
  build:
    docker:
      - image: circleci/node:14
    steps:
      - checkout
      - run:
          command: echo "This is working"
```

Let's quickly go through all the steps in the file, as follows:

1. The file starts with a `version` specification. We are using version 2.1, as this is the current version of CircleCI.

2. Then, we will have a list of `jobs` that get executed in parallel. As we only have one thing that we want to do, we can only see the `build` job that we are running. Later, we will add the whole Docker build and publish the functionality here.

3. Each job receives an executor type, which needs to be `machine`, `docker`, or `macos`. We are using the `docker` type because we can rely on many prebuilt images of CircleCI. The image is specified in a separate `image` property. There, I have specified `node` in version 14, because we need Node.js for our CI workflow.

4. Each job then receives several steps that are executed with every commit that is pushed to the Git repository.

5. The first step is the `checkout` command, which clones the current version of our repository so that we can use it in any further steps.

6. Lastly, to test that everything has worked, we use the `run` step. It lets us execute a command directly in the Docker `node:14` image that we have started with CircleCI. Each command that you want to execute must be prefixed with `command`.

The result of this config file should be that we have pulled the current master branch of our application and printed the text This is working at the end. To test the CircleCI setup, commit and push this file to your GitHub or Bitbucket repository.

CircleCI should automatically notify you that it has started a new **CI** job for our repository. You can find the job by hitting the **Jobs** button in the left-hand panel of CircleCI. The newest job should be at the top of the list. Click on the job to see the details. They should look as follows:

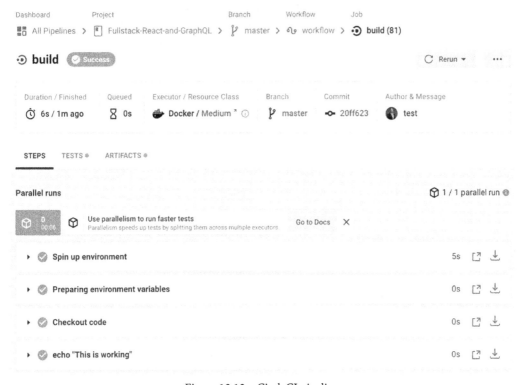

Figure 12.12 – CircleCI pipeline

In the preceding screenshot, each step is represented in a separate row at the bottom of the window. You can expand each row to see the logs that are printed while executing the specific command shown in the current row. The preceding screenshot shows that the job has been successful.

Now that we have configured CircleCI to process our repository on each push, we must take a look at how to host and deploy our application directly, after finishing the build.

Deploying applications to Amazon ECS

CircleCI executes our build steps each time we push a new commit. Now, we want to build our Docker image and deploy it automatically to a machine that will serve our application to the public.

Our database and uploaded images are hosted on AWS already, so we can also use AWS to serve our application. Setting up AWS correctly is a significant task, and it takes a large amount of time. We will use Amazon ECS to run our Docker image. Still, to correctly set up the network, security, and container registry is too complex to be explained in just one chapter. I recommend that you take a course or pick up a separate book to understand and learn advanced setups with AWS, and the configuration that is needed to get production-ready hosting. For now, we will use ECS to get the container, including the database connection, running.

Before directly going to Amazon ECS and creating your cluster, we need to prepare two services – one is AWS **ALB**, which stands for **Application Load Balancer**, and the other is Amazon ECR. If you set up an ECS cluster, there will be one or more instances of the same service or, to be exact, task running. Those tasks need to receive traffic, but if you release a new version, they should also be exchanged with new tasks while still serving the traffic. That is a good job for AWS ALB, as it can split traffic between the instances and handle dynamic port mapping if tasks are exchanged.

Log in to the AWS Management Console, search for the EC2 service, and click on it. Then, follow these instructions:

1. Scroll down on the left panel until you see the **Load Balancing** section, as shown in the following screenshot:

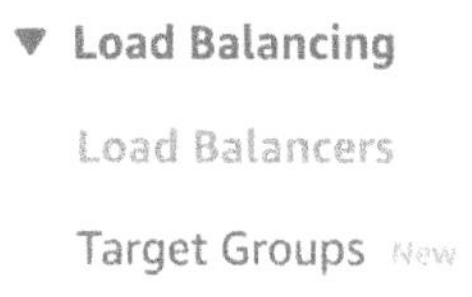

▼ **Load Balancing**

Load Balancers

Target Groups New

Figure 12.13 – AWS Load Balancing section

2. Click on **Load Balancers** and you will see an empty list. We need to click on **Create Load Balancer** in the top left corner now. On the next page, select **Application Load Balancer**.

3. We need to specify a name for our load balancer and a scheme. We will go with the **Internet-facing** option because that way we can access it from outside AWS:

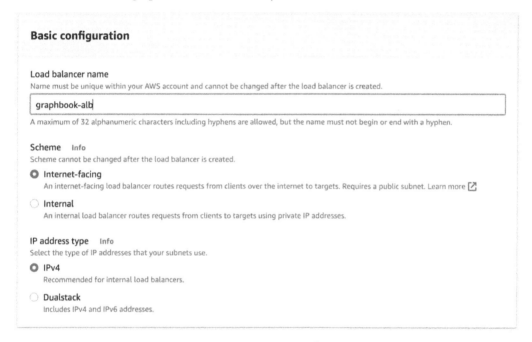

Figure 12.14 – AWS ALB configuration

Important Note

Normally, you would not make the load balancer public but instead add another **Content Delivery Network (CDN)**, cache, and firewall in front of your application to protect it from **Distributed Denial of Service (DDoS)** attacks or unnecessary load. Services on AWS such as Route53 and CloudFront work well together to accomplish this, but they are out of the scope of this book.

4. In the next step, we need to select a VPC that will bring the attached resources into one private network. There should be an existing one from the database that you can select. Please do that and select the availability zones, as shown in the following screenshot:

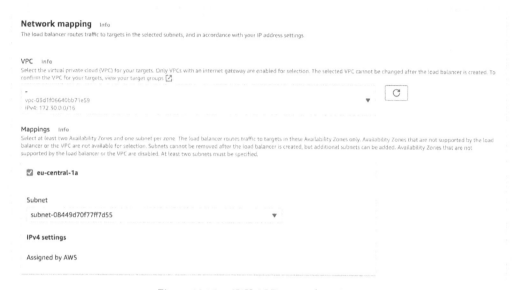

Figure 12.15 – AWS ALB network settings

5. Next, you need to select the security group that we have created with the AWS RDS database, as shown in the following screenshot:

Figure 12.16 – AWS ALB Security groups

6. If you scroll further down, you can see that you need to specify the routing for your load balancer, that is, where the load or traffic needs to go, depending on some rules that you need to define. To do that, we need to define a target group. There should be a link reading **Create target group** below the selection input, as shown in the following screenshot:

Figure 12.17 – AWS ALB routing

If you have created and selected the target group, it should look like the preceding figure, so let's do it.

7. Open the link in a separate tab or window; as mentioned before, you should see the following screen:

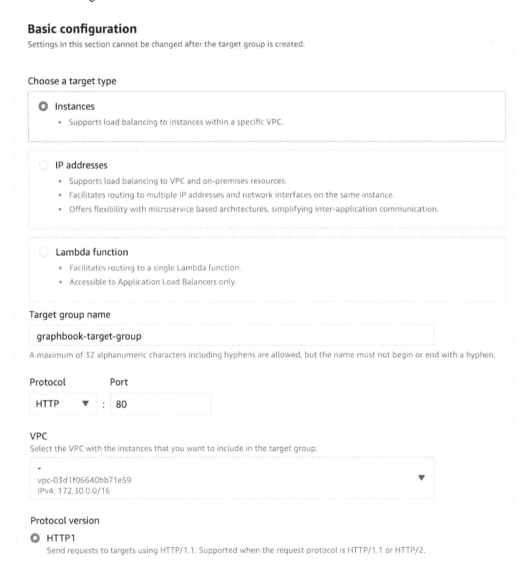

Figure 12.18 – AWS target group creation

8. Select **Instances,** which means load balancing between different instances in one VPC. Give the target group a name and select the same VPC as before.

9. After that, you can hit the **Next** button. You will be presented with the following screen:

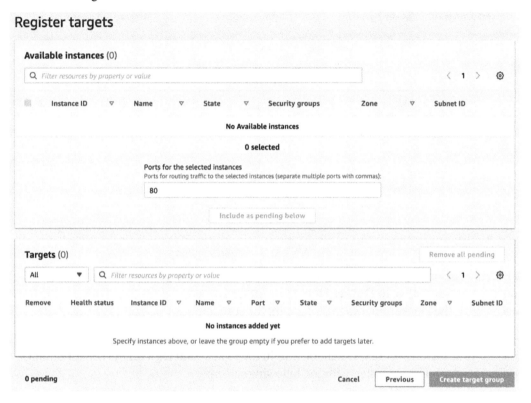

Figure 12.19 – AWS target group targets

This screen normally shows you all instances that would be included in your target group. Because we did not create the ECS cluster yet, this is empty. You can continue by hitting the **Create target group** button.

10. Go back to the wizard to set up AWS ALB. Hit the **Refresh** button next to the target group selection and select the target group.

11. Hit the **Create load balancer** button at the end of the screen.

Finally, you should have reached the end of the wizard and be presented with a summary, as shown in the following screenshot:

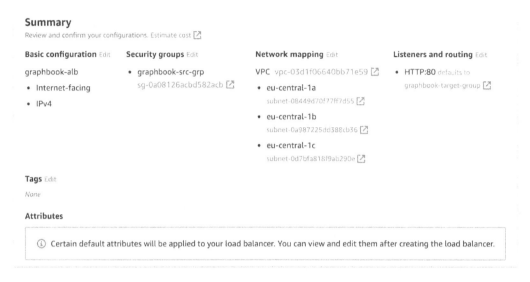

Figure 12.20 – AWS ALB creation summary

The next thing we need to prepare is an Amazon ECR repository. Amazon ECR is nothing more than an alternative to Docker Hub or any other Docker registry. In a Docker registry, you can push the Docker images that you built for your application. This is the basis on which our ECS cluster will run.

To set up your ECR repository, search for ECR in the top bar and click on the corresponding service. You should be presented with the following screen:

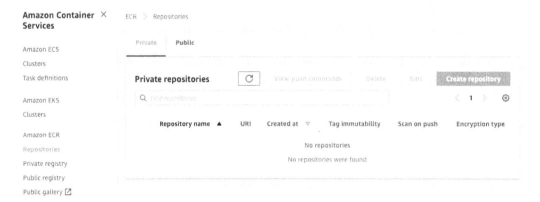

Figure 12.21 – Amazon ECR overview

To set up your Amazon ECR repository, follow these steps:

1. Click on **Create repository** on the right side to get into the creation wizard.

2. Next, you need to provide a name for your repository, as shown in the following screenshot:

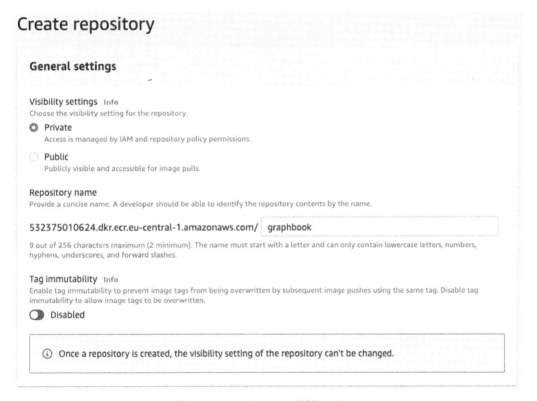

Figure 12.22 – Amazon ECR settings

We leave our registry private because no one external should have access to it. If you need to have it public for everyone, you can change this setting.

3.   Click **Create repository** to set it up. You will see the following update table now:

Figure 12.23 – Amazon ECR repository created

4.   Later, you will need the URI, which is shown next to the registry name.

Now we are prepared to start setting up the ECS cluster. To find ECS, just go to the services bar at the top and search for ECS. If you click this service and go to the **Clusters** section, it will show you all the running ECS clusters. It should look like the following:

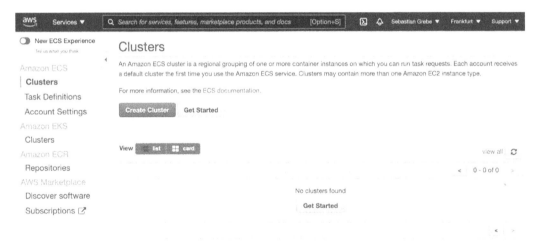

Figure 12.24 – Amazon ECS Clusters

The process to configure ECS is very complex, and we will follow the most basic configuration in the scope of this book. Follow these instructions to get it working:

1. Hit the **Create Cluster** button (see *Figure 12.24*) to get started.

2. On the next screen, you will be asked what kind of instances you want to use. We are going to use the EC2 Linux instances, as shown in the following screenshot:

Figure 12.25 – Amazon ECS cluster template

3. Click **Next** to get to the configuration wizard for your cluster.

4. You will be asked for a cluster name. The default configuration is fine otherwise. The only option that requires a change is the instance type. I recommend going with `t2.micro` because it does not cost that much, which is good for development. The **Number of instances** option specifies how many parallel running EC2 instances we want. For development, mostly one is fine, but if you need to scale, this is something you need to increase:

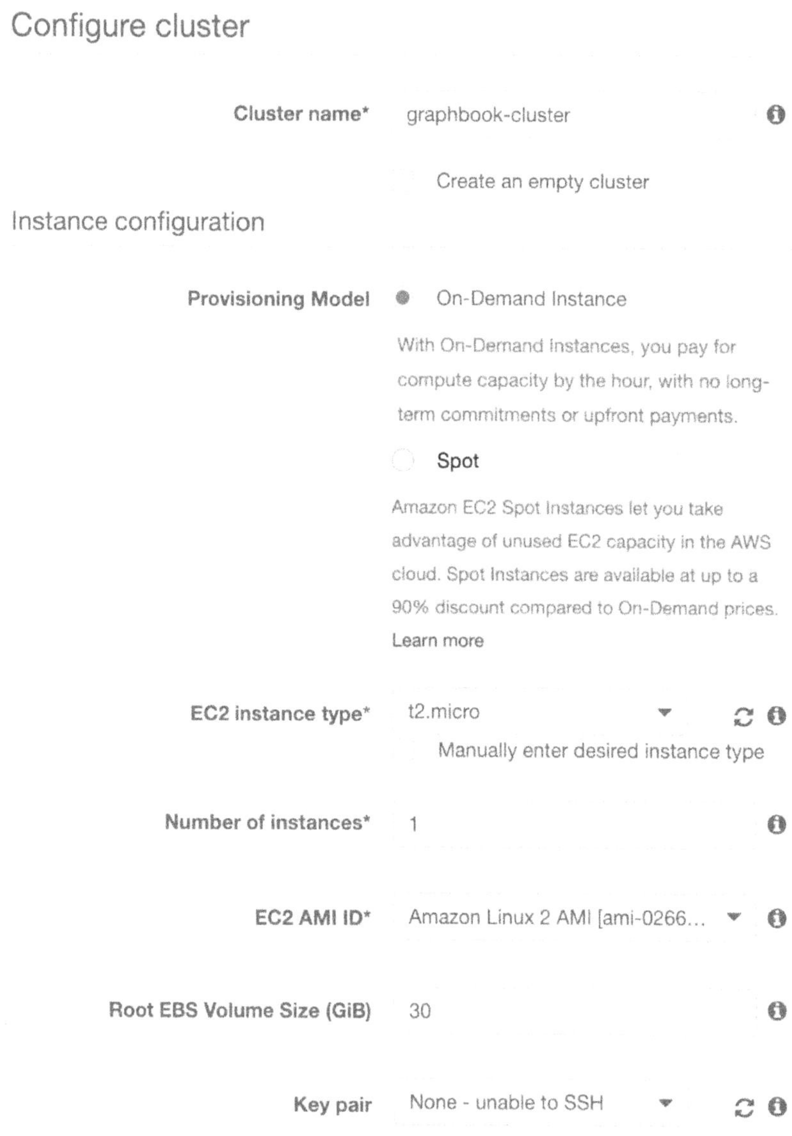

Configure cluster

Cluster name*	graphbook-cluster ⓘ
	Create an empty cluster

Instance configuration

Provisioning Model
   ●   On-Demand Instance

With On-Demand Instances, you pay for compute capacity by the hour, with no long-term commitments or upfront payments.

   ○   **Spot**

Amazon EC2 Spot Instances let you take advantage of unused EC2 capacity in the AWS cloud. Spot Instances are available at up to a 90% discount compared to On-Demand prices. Learn more

EC2 instance type*	t2.micro ▼ ⟳ ⓘ
	Manually enter desired instance type
Number of instances*	1 ⓘ
EC2 AMI ID*	Amazon Linux 2 AMI [ami-0266... ▼ ⓘ
Root EBS Volume Size (GiB)	30 ⓘ
Key pair	None - unable to SSH ▼ ⟳ ⓘ

Figure 12.26 – Amazon ECS cluster configuration

5. Scroll down to provide some further networking configuration. As we have selected three subnets when configuring the load balancer, we should also now select those three subnets:

Networking

Configure the VPC for your container instances to use. A VPC is an isolated portion of the AWS cloud populated by AWS objects, such as Amazon EC2 instances. You can choose an existing VPC, or create a new one with this wizard.

Figure 12.27 – Amazon ECS cluster network settings

6. After selecting the subnets, you need to select the security group that we also used for the ALB configuration:

Figure 12.28 – Amazon ECS Security settings

7. Hit **Create Cluster** and AWS will start the process to spin everything up. This might take some time.

8. Once AWS is done, you can click the **View Cluster** button, which will take you to the detailed cluster page.

We have defined now that AWS is running an ECS cluster based on one EC2 instance. One thing we did not do so far is to define what this cluster does. For that, we need to go to **Task definitions** on the left-hand panel. Then, follow these instructions:

1. Click on the **Create new task definition** button.

2. Select the **EC2** type to make your task compatible with this type of cluster.

3. Give your task definition a name and the **ecsTaskExecutionRole** role. It should look like the following screenshot:

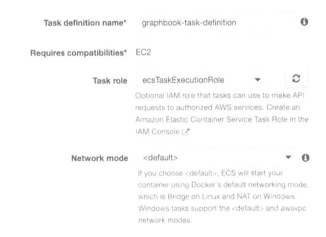

Configure task and container definitions

A task definition specifies which containers are included in your task and how they interact with each other. You can also specify data volumes for your containers to use. Learn more

Task definition name*   graphbook-task-definition

Requires compatibilities*   EC2

Task role   ecsTaskExecutionRole
Optional IAM role that tasks can use to make API requests to authorized AWS services. Create an Amazon Elastic Container Service Task Role in the IAM Console

Network mode   <default>
If you choose <default>, ECS will start your container using Docker's default networking mode, which is Bridge on Linux and NAT on Windows. Windows tasks support the <default> and awsvpc network modes.

Task execution IAM role

This role is required by tasks to pull container images and publish container logs to Amazon CloudWatch on your behalf. If you do not have the ecsTaskExecutionRole already, we can create one for you.

Task execution role   ecsTaskExecutionRole

Figure 12.29 – Amazon ECS task definition

4. Scroll down and give your task a size, which means a memory size and CPU size that it will need to process its task. It should look like the following screenshot:

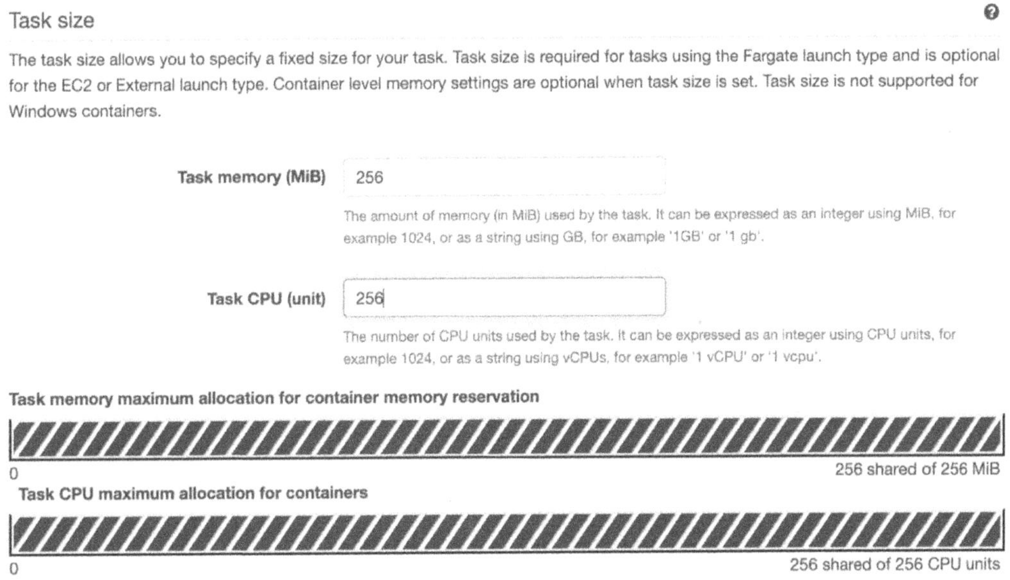

Task size   ❓

The task size allows you to specify a fixed size for your task. Task size is required for tasks using the Fargate launch type and is optional for the EC2 or External launch type. Container level memory settings are optional when task size is set. Task size is not supported for Windows containers.

Task memory (MiB) | 256 |

The amount of memory (in MiB) used by the task. It can be expressed as an integer using MiB, for example 1024, or as a string using GB, for example '1GB' or '1 gb'.

Task CPU (unit) | 256 |

The number of CPU units used by the task. It can be expressed as an integer using CPU units, for example 1024, or as a string using vCPUs, for example '1 vCPU' or '1 vcpu'.

Task memory maximum allocation for container memory reservation

0   256 shared of 256 MiB

Task CPU maximum allocation for containers

0   256 shared of 256 CPU units

Figure 12.30 – Amazon ECS task definition size

This setting needs to be aligned with the resources your cluster has.

5. Now, click the **Add container** button to define which types of containers will run inside of this ECS task:

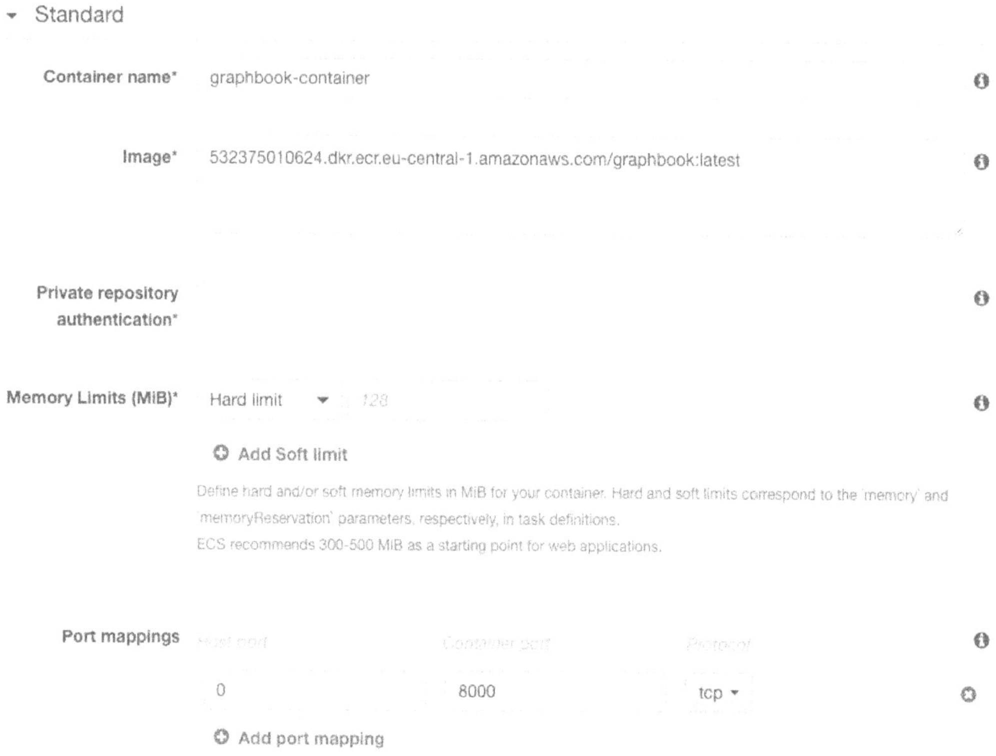

▾ Standard

Container name*	graphbook-container	🛈
Image*	532375010624.dkr.ecr.eu-central-1.amazonaws.com/graphbook:latest	🛈
Private repository authentication*		🛈
Memory Limits (MiB)*	Hard limit ▾ 128	🛈

    ⊕ Add Soft limit

Define hard and/or soft memory limits in MiB for your container. Hard and soft limits correspond to the 'memory' and 'memoryReservation' parameters, respectively, in task definitions.
ECS recommends 300-500 MiB as a starting point for web applications.

Port mappings	Host port	Container port	Protocol	🛈
	0	8000	tcp ▾	⊗

    ⊕ Add port mapping

Figure 12.31 – Amazon ECS task definition container configuration

In this dialog, you first need to provide a name for your container. Secondly, you need to provide the URI from the ECR registry that we created at the beginning. You need to append the tag for your image, which will be `:latest` for the moment. As we did not publish any image yet, it will not work anyway, but we will fix this later.

6. Scroll down to the **Port mappings** section. Fill **Container port** with the number `8000` because this is the default port that we use for Graphbook. **Host port** needs to stay at `0`, as this will be automatically assigned by your load balancer. AWS ALB will dynamically map a free port to the container port. We do not need to care exactly which port it will be.

7. Under the environment section, we need to add all environment variables that our application requires to start. The variables shown in the following screenshot should be enough to get your container running:

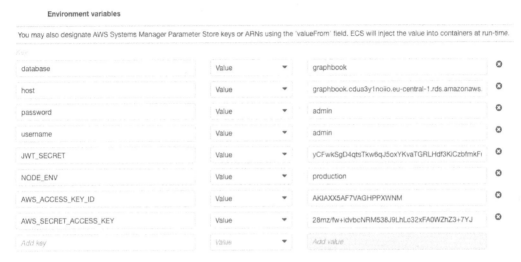

Figure 12.32 – Amazon ECS task definition container environment variables

8. The only setting that is helpful is under the **Storage and logging** section. I recommend activating CloudWatch Logs, as you can then see all the logs from your application. It should look like the following screenshot:

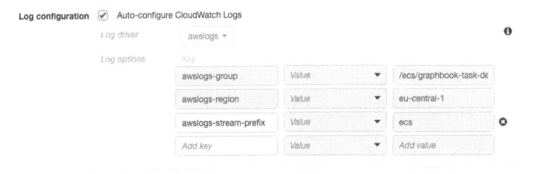

Figure 12.33 – Amazon ECS task definition container logs

9. After enabling this option, you can hit the **Add** button at the bottom of the dialog.

These are all the important things that you need to do. There is a multitude of further detailed configurations that you can do. For us, they are not required to get our application running, and they are out of the scope of this book.

The container definitions table should look like the following screenshot:

Figure 12.34 – Amazon ECS task definition container definitions

What we just did is the most basic ECS setup that you can do. Across all the services that we just set up, we used the simplest configuration that you can do, but there is a tremendous amount of advanced setups and configurations we did not have a look at.

Now, you can click on the **Create** button at the very end of the ECS task definition wizard. After AWS has successfully created the task definition, go back to the main **Clusters** screen of ECS. You will see this cluster overview:

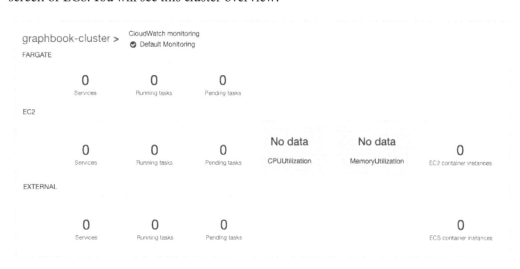

Figure 12.35 – Amazon ECS cluster overview

It just shows you that, still, nothing is running inside your cluster. To fix that, click on your cluster name at the top left.

The actual problem is that no service has been created that is using the just-configured task definition, as shown in the following screenshot:

Figure 12.36 – Amazon ECS cluster services

Hit the **Create** button above the table.

Then, you need to provide the following data:

1. For **Launch type**, you need to select **EC2**, as with the previous steps.

2. Then, you need to select the task definition family, which should match the name of your previously created task definition.

3. Also, you need to select the cluster that we are currently looking at.

4. You need to provide a service name for your service, such as `graphbook-service`.

5. For the **Number of tasks** option, the value `1` is fine. It means that only one task will run in this service at the same time. This is also restricted by the health percentage. The minimum health percent of `100` means that at least one service that is correctly functioning should be running, whereas the maximum percent of `200` means that only a maximum of two tasks that are healthy are allowed to be running. This restriction is required for the moment where you update your service with a new application version. At this moment, there will be two versions running at the same time that will be exchanged. So, currently, we have `200` health percent.

That is all the information that you need to provide. The rest of the settings can be left as they are set by default. The result should look like the following screenshot:

Figure 12.37 – Amazon ECS service configuration

You can now hit the **Next step** button at the bottom of the screen to continue.

The next screen is very important because this requires our AWS ALB to be set up.

You need to select the **Application Load Balancer** option and then select our previously created ALB. The settings should match the configuration in the following screenshot:

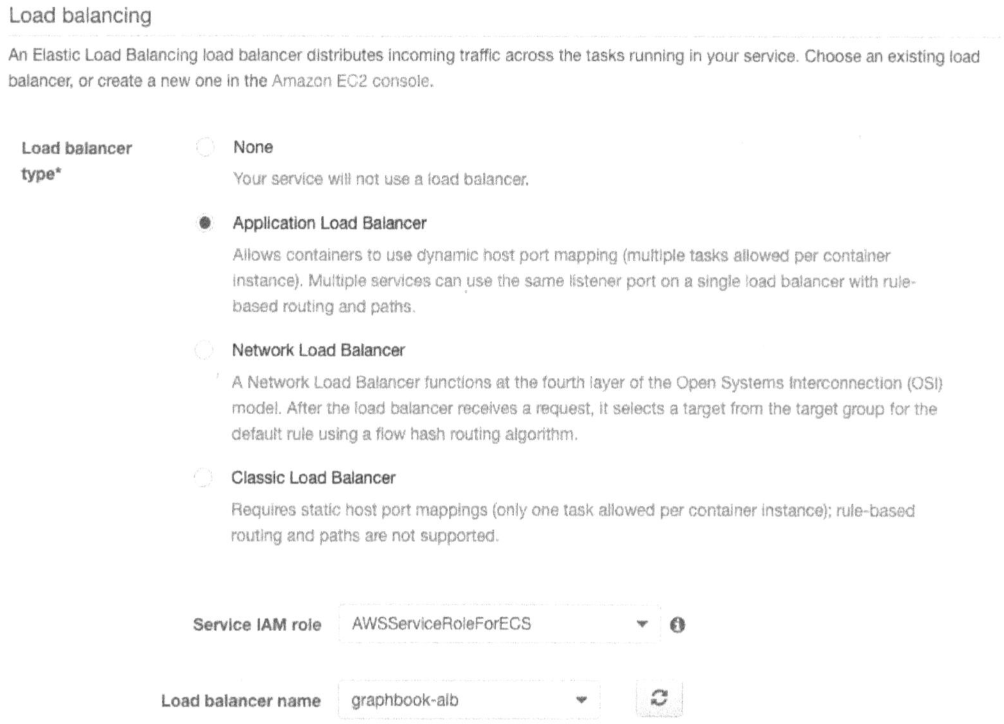

Load balancing

An Elastic Load Balancing load balancer distributes incoming traffic across the tasks running in your service. Choose an existing load balancer, or create a new one in the Amazon EC2 console.

Load balancer
type*

○ None
  Your service will not use a load balancer.

● Application Load Balancer
  Allows containers to use dynamic host port mapping (multiple tasks allowed per container instance). Multiple services can use the same listener port on a single load balancer with rule-based routing and paths.

○ Network Load Balancer
  A Network Load Balancer functions at the fourth layer of the Open Systems Interconnection (OSI) model. After the load balancer receives a request, it selects a target from the target group for the default rule using a flow hash routing algorithm.

○ Classic Load Balancer
  Requires static host port mappings (only one task allowed per container instance); rule-based routing and paths are not supported.

Service IAM role | AWSServiceRoleForECS ▼ | ❶

Load balancer name | graphbook-alb ▼ | ↺

Figure 12.38 – Amazon ECS service load balancing

This setting will make use of our ALB and do the dynamic port mapping to the container that is running within this service's tasks.

If we scroll down, we need to provide details on how the ALB will map to the container. Now, you should see this message:

Container to load balance

Container name : port | graphbook-container:0:8000 ▼ | Add to load balancer

Figure 12.39 – Amazon ECS service load balancing mapping

You need to click the **Add to load balancer** button. The good thing is you can just select the target group that we created previously. After selecting it, you should see the following screen:

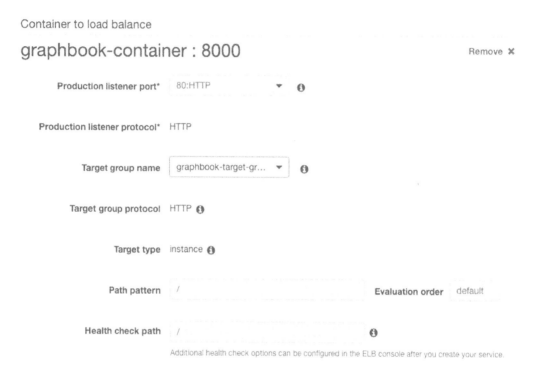

Figure 12.40 – Amazon ECS load balancing target group configuration

You can now hit the **Next step** button at the end of the screen and on the next screen where it asks for auto-scaling, which we do not require. You should be able to click the **Create Service** option at the bottom of the summary screen.

AWS will try to create the service now and spin up the tasks. The problem is that we did not push any Docker image so far. The ECS service will not be able to spin up any task correctly because of that.

So, let's fix that.

Setting up CircleCI with Amazon ECR and ECS

We will start with a blank CircleCI config again; so, empty the old `config.yml` file.

One important thing we did not do so far is to set up automated testing within our pipeline. Otherwise, our commits will trigger an automated pipeline and just deploy the untested code, which might cause production issues that we want to prohibit.

So, let's do this first. Follow these steps:

1. Insert these lines into our `config.yml` file, as follows:

```yaml
version: 2.1
jobs:
  test:
    docker:
      - image: circleci/node:14
        auth:
          username: $DOCKERHUB_USERNAME
          password: $DOCKERHUB_PASSWORD
        environment:
          host: localhost
          username: admin
          password: passw0rd
          database: graphbook
          JWT_SECRET: 1234
```

This configuration creates a `test` job and pulls one Docker image. The Docker image is the Node.js image from CircleCI that we will run our application inside for testing purposes. At the same time, we pass credentials to actually pull the image, but we also pass some default environment variables that we will make use of in the next step.

2. Add another image to the `test` job:

```yaml
      - image: circleci/mysql:8.0.4
        command: [--default-authentication-
                  plugin=mysql_native_password]
        auth:
          username: $DOCKERHUB_USERNAME
          password: $DOCKERHUB_PASSWORD
        environment:
```

```
MYSQL_ROOT_PASSWORD: passw0rd
MYSQL_DATABASE: graphbook
MYSQL_USER: admin
MYSQL_PASSWORD: passw0rd
```

This image is for the MySQL database, against which we can run our migrations but also test scripts. This will be created from scratch every time the pipeline runs. You can see here that we also provide the same environment variables. This will set up the MySQL database with these credentials and the Node.js container will have those credentials in the environment variables.

3. As you can see in the preceding steps, we are using syntax such as $DOCKERHUB_ USERNAME to inject variables from the CircleCI settings into our pipeline. That way, we do not need to repeat them repeatedly, but also, they are not committed into our code. Set up the environment variables according to the following screenshot:

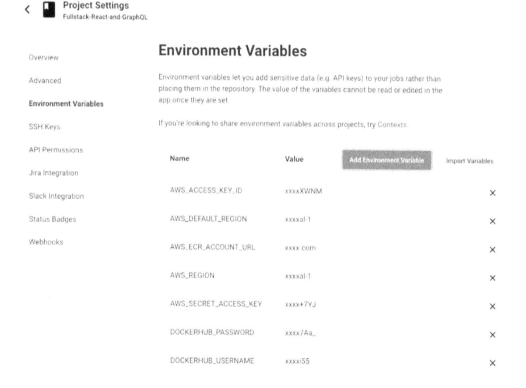

Figure 12.41 – CircleCI project environment variables

4. Now, we need to add functionality to actually run our tests. We will do this inside a `steps` property that CircleCI is able to understand. Just add the code below the previously added `jobs` section:

```
steps:
  - checkout
  - run: npm install
```

The flow of the test job is quite easy. First, we check out our code and then we install all dependencies that our application requires.

5. Then, we also require the same Sequelize packages to run our database migrations as we had in our `Dockerfile`. Add the following code to do so:

```
- run:
    name: "Install Sequelize"
    command: sudo npm install -g mysql2
            sequelize sequelize-cli
```

6. Then, we need to wait for the database image to come up. If we do not do that, when taking the next steps, the commands will fail if the database is not yet running. Add the following code to wait for our database to come up:

```
- run:
    name: Waiting for MySQL to be ready
    command: |
      for i in 'seq 1 10';
        do
            nc -z 127.0.0.1 3306 && echo Success
            && exit 0
          echo -n .
          sleep 1
        done
        echo Failed waiting for MySQL && exit 1
```

7. Once the database is up, we can run our database migrations against the `test` database. Add the following code to run them:

```
- run:
    name: "Run migrations for test DB"
    command: sequelize db:migrate
      --migrations-path src/server/migrations
      --config src/server/config/index.js
      --env production
```

8. Now, we can finally run our test against the freshly created database. Just use the below code to get it working:

```
- run:
    name: "Run tests"
    command: npm run test
    environment:
      NODE_ENV: production
```

If one of the tests fails, the complete pipeline will fail. This ensures that only working application code is released to the public.

9. The last thing to do to get our automated tests running is to set up the CircleCI workflow. You can copy the following code to get it running:

```
workflows:
  build-and-deploy:
    jobs:
      - test
```

You can commit and push this new config file into your Git repository, and CircleCI should automatically process it and create a new job for you.

The resulting job should look as follows:

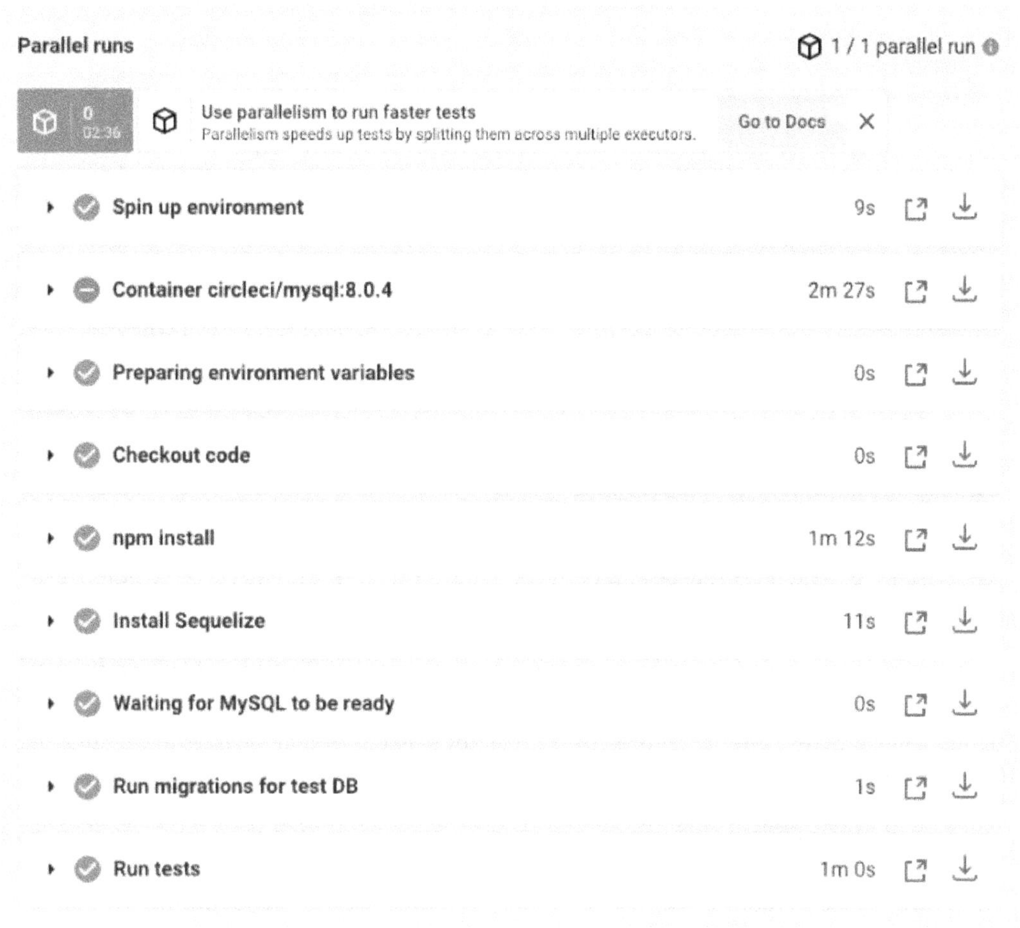

Figure 12.42 – CircleCI test pipeline

Next, we need to build the Docker image and push this to our registry. Thanks to CircleCI, this is quite easy.

Add this configuration to your `config.yml` file below the version specification:

```
orbs:
  aws-ecr: circleci/aws-ecr@7.2.0
```

A CircleCI orb is a set of package configurations that you can share or make use of without needing to write all the steps on your own. This orb that we have just added can build and push a Docker image to Amazon ECR, which we set up in the previous section.

> **Important Note**
> You can find all the available CircleCI orbs and their documentation on the
> official CircleCI orb website: `https://circleci.com/developer/`
> `orbs`.

To leverage this orb, add one workflow step, as shown in the following screenshot:

```
- aws-ecr/build-and-push-image:
  repo: "graphbook"
  tag: "${CIRCLE_SHA1}"
  requires:
    - test
```

The preceding configuration will build and push your Docker image to Amazon ECR, specified by the `repo` attribute. It will also wait for the `test` step because we have mentioned this in the `requires` property.

If you commit and push this configuration, you will see that in the CircleCI pipeline, there is a separate ECR step. If that is completed, you will be able to find a new Docker image inside the ECR repository.

The only thing missing now is to make use of this Docker image inside of Amazon ECS. If you remember, we specified the Docker image inside of our Amazon ECS task definition. Updating this manually after each pipeline run is not feasible though. To automate this process, add one further orb at the top of the CircleCI config:

```
aws-ecs: circleci/aws-ecs@02.2.1
```

CircleCI also has us covered if we want to update and push a new task definition to our service. To leverage this orb, add this code as the last workflow step:

```
- aws-ecs/deploy-service-update:
  requires:
    - aws-ecr/build-and-push-image
    - test
  family: "graphbook-task-definition"
  cluster-name: "graphbook-cluster"
  service-name: "graphbook-service"
  container-image-name-updates: "container=
    graphbook-container,tag=${CIRCLE_SHA1}"
```

This step waits for both the test and the ECR job to complete. Afterward, it will create a new task definition revision on Amazon ECS with the given `family` name. It will then update the service with the given name inside the given cluster.

Commit and push the new config file, and you will see the following pipeline with three jobs running:

Figure 12.43 – CircleCI CD pipeline

Amazon ECS will take some time to replace the currently running task, but after that, the new version of your application will be running.

Still, the question is how we can access Graphbook now. For that, we can go to our AWS ALB, go to the **Load Balancers** section, and click our ALB. It will show the following information:

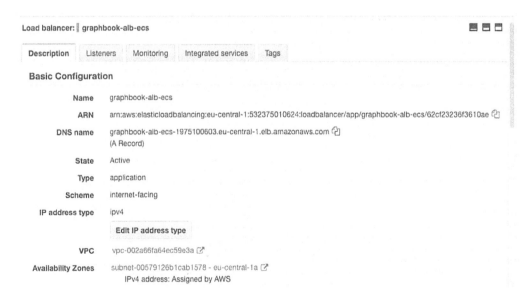

Figure 12.44 – AWS ALB DNS name

Under the given DNS name, we can access the load balancer and, via that, our application.

As mentioned previously, this is not recommended, but the fully fledged AWS setup is out of the scope of this book. You should be able to access Graphbook under that link.

That is all you need to do. It will test the application code with our test suite, build and push a new Docker image, and lastly update the task definition and ECS service to replace the old task with an updated task with the new task definition.

Summary

In this chapter, you learned how to dockerize your application using a normal `Dockerfile` and a multi-stage build.

Furthermore, I have shown you how to set up an exemplary CD workflow using CircleCI and AWS. You can replace the deployment process with a more complex setup while continuing to use your Docker image.

Having read this chapter, you have learned everything from developing a complete application to deploying it to a production environment. Your application should now be running on Amazon ECS.

At this point, you have learned all the important things, including setting up React with Webpack, developing a local setup, server-side rendering, and how to tie all the things together with GraphQL. Also, you are able to release changes frequently with CD. Looking ahead, there are still things that we can improve – for example, the scalability of our application or bundle splitting, which are not handled in the scope of this book, but there are many resources available that will help you improve in these areas.

I hope you enjoyed the book and wish you every success!

Index

Packt.com

Subscribe to our online digital library for full access to over 7,000 books and videos, as well as industry leading tools to help you plan your personal development and advance your career. For more information, please visit our website.

Why subscribe?

- Spend less time learning and more time coding with practical eBooks and Videos from over 4,000 industry professionals

- Improve your learning with Skill Plans built especially for you

- Get a free eBook or video every month

- Fully searchable for easy access to vital information

- Copy and paste, print, and bookmark content

Did you know that Packt offers eBook versions of every book published, with PDF and ePub files available? You can upgrade to the eBook version at packt.com and as a print book customer, you are entitled to a discount on the eBook copy. Get in touch with us at customercare@packtpub.com for more details.

At www.packt.com, you can also read a collection of free technical articles, sign up for a range of free newsletters, and receive exclusive discounts and offers on Packt books and eBooks.

Other Books You May Enjoy

If you enjoyed this book, you may be interested in these other books by Packt:

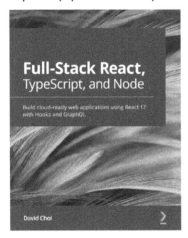

Full-Stack React, TypeScript, and Node

David Choi

ISBN: 978-1-83921-993-1

- Discover TypeScript's most important features and how they can be used to improve code quality and maintainability
- Understand what React Hooks are and how to build React apps using them
- Implement state management for your React app using Redux
- Set up an Express project with TypeScript and GraphQL from scratch
- Build a fully functional online forum app using React and GraphQL
- Add authentication to your web app using Redis
- Save and retrieve data from a Postgres database using TypeORM
- Configure NGINX on the AWS cloud to deploy and serve your apps

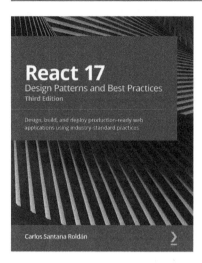

React 17 Design Patterns and Best Practices - Third Edition

Carlos Santana Roldán

ISBN: 978-1-80056-044-4

- Get to grips with the techniques of styling and optimizing React components
- Create components using the new React Hooks
- Get to grips with the new React Suspense technique and using GraphQL in your projects
- Use server-side rendering to make applications load faster
- Write a comprehensive set of tests to create robust and maintainable code
- Build high-performing applications by optimizing components

Packt is searching for authors like you

If you're interested in becoming an author for Packt, please visit authors. packtpub.com and apply today. We have worked with thousands of developers and tech professionals, just like you, to help them share their insight with the global tech community. You can make a general application, apply for a specific hot topic that we are recruiting an author for, or submit your own idea.

Hi!

I am Sebastian Grebe, author of *Full-Stack Web Development with GraphQL and React Second Edition*. I really hope you enjoyed reading this book and found it useful for increasing your productivity and efficiency in Web Development.

It would really help me (and other potential readers!) if you could leave a review on Amazon sharing your thoughts on *Full-Stack Web Development with GraphQL and React Second Edition*.

Go to the link below or scan the QR code to leave your review:

`https://packt.link/r/1801077886`

Your review will help me to understand what's worked well in this book, and what could be improved upon for future editions, so it really is appreciated.

Best Wishes,